Engineering Neural Tissue from Stem Cells

Engineering Neural Tissue from Stem Cells

Stephanie Willerth

Academic Press is an imprint of Elsevier
125 London Wall, London EC2Y 5AS, United Kingdom
525 B Street, Suite 1800, San Diego, CA 92101-4495, United States
50 Hampshire Street, 5th Floor, Cambridge, MA 02139, United States
The Boulevard, Langford Lane, Kidlington, Oxford OX5 1GB, United Kingdom

Notices
Knowledge and best practice in this field are constantly changing. As new research and experience broaden our understanding, changes in research methods, professional practices, or medical treatment may become necessary.

Practitioners and researchers must always rely on their own experience and knowledge in evaluating and using any information, methods, compounds, or experiments described herein. In using such information or methods they should be mindful of their own safety and the safety of others, including parties for whom they have a professional responsibility.

To the fullest extent of the law, neither the Publisher nor the authors, contributors, or editors, assume any liability for any injury and/or damage to persons or property as a matter of products liability, negligence or otherwise, or from any use or operation of any methods, products, instructions, or ideas contained in the material herein.

Library of Congress Cataloging-in-Publication Data
A catalog record for this book is available from the Library of Congress

British Library Cataloguing-in-Publication Data
A catalogue record for this book is available from the British Library

ISBN 978-0-12-811385-1

For information on all Academic Press publications
visit our website at https://www.elsevier.com/books-and-journals

Publisher: Mara Conner
Acquisition Editor: Chris Katsaropoulos
Editorial Project Manager: Anna Valutkevich
Production Project Manager: Sruthi Satheesh
Cover Designer: Mark Rogers

Typeset by SPi Global, India

Contents

Author Biography

Stephanie Willerth currently holds a Canada Research Chair in Biomedical Engineering at the University of Victoria where she is dually appointed in the Department of Mechanical Engineering and Division of Medical Sciences as an Associate Professor. Her interdisciplinary research group investigates how to engineer neural tissue by combining pluripotent stem cells, controlled drug delivery, and biomaterial scaffolds. Her honors include being named a Women of Innovation in 2017, a 2015 Young Innovator in Cellular and Molecular Biology, a Star in Global Health by Grand Challenges Canada, and the 2014 Faculty of Engineering Award for Excellence in Teaching. She is an active member of the Stem Cell Network and the International Collaboration on Repair Discoveries, who supported her 2016 sabbatical leave at the Wisconsin Institute for Discovery where she authored this book. She served as both the Director of the Centre for Biomedical Research and the President of the Canadian Biomaterials Society during 2017. Before accepting her faculty position, Willerth completed a National Institutes of Health sponsored postdoctoral fellowship at the University of California—Berkeley and graduate studies at Washington University. She received undergraduate degrees in Biology and Chemical Engineering from the Massachusetts Institute of Technology.

Foreword

Dr. Willerth has brought together the main themes underlying tissue engineering in the central nervous system, with a focus on the brain, the spinal cord and including both traumatic and degenerative diseases. The book follows a natural progression: the first three chapters set the stage with a description of the central nervous system, its diseases and disorders, and the stem cells used therein; the subsequent three chapters, 4–6, focus on the role of biomaterials and tissue engineering in overcoming these diseases in conjunction with stem cells; the last two chapters, 7 and 8, highlight the benefit of therapeutics co-delivered with stem cells and an outlook to the future.

In what many consider the final frontier of medicine, the central nervous system has been a formidable challenge, with a series of biological barriers making access and regeneration more difficult. It is important to realize that there is often very little that can be done for patients suffering from traumatic injuries such as stroke and spinal cord injury. For stroke, tissue plasminogen activator is the only approved treatment and it must be administered within 4 h to the patient. In spinal cord injury, methylprednisolone is administered to patients, but its benefit has been questioned. Importantly, a recent clinical trial led by Asterias has shown very promising results in spinal cord injury with the transplantation of oligodendrocyte progenitor cells derived from embryonic stem cells. There is also excitement about ongoing clinical trials for degenerative diseases like blindness due to age-related macular degeneration.

Whether transplantation of exogenous stem cells or stimulation of endogenous stem cells, biomaterials can have a profound impact on success. In both cases, cell survival and integration are key barriers. Biomaterials can provide the environment that promotes greater cell survival by manipulating the mechanical and biochemical properties. Similarly, biomaterials can be used to deliver biomolecules locally to the brain or spinal cord, to stimulate endogenous stem cells and/or promote integration by breaking down barriers, such as that formed by the glial scar. With this book, Dr. Willerth captures some of the key challenges and opportunities that face engineering neural tissue in the central nervous system with stem cells.

Molly S. Shoichet
University of Toronto

Acknowledgments

It takes a lot of time and support to write a book, here is where I attempt to compile the list of people who supported me throughout this project and show my appreciation.

First, I must thank everyone at Elsevier for their support, including my editors, Anna Valutkevich and Chris Katsaropoulos, along with my permissions team, Narmatha Mohan, Vinoth Kumar Balu, and the production staff.

I am thankful for the members of Willerth lab past and present for inspiring this work. I would like to highlight Meghan Robinson, Krista Wilson, Michaela Thomas, Laura de la Vega, and Andrew Agbay for their constructive feedback on the contents of this book.

My research group has received funding from various agencies, enabling my group to conduct neural tissue engineering research. This research, in turn, served as an inspiration for this project. These funding sources include the Canada Research Chairs program, the Natural Sciences and Engineering Resource Council, the Stem Cell Network, the Rick Hansen Institute, and Grand Challenges Canada. Additional support for this project came from the International Collaboration on Repair Discoveries and the Rick Hansen Institute, which sponsored my sabbatical leave at the University of Wisconsin through their International Exchange Award program. I would like to thank the Wisconsin Institute for Discovery at the University of Wisconsin—Madison for hosting me with special thanks to Randolph Ashton and Krishanu Saha. It provided an inspiring environment where I wrote majority of these books. I would like to thank my colleagues at the University of Victoria in the Department of Mechanical Engineering and in the Division of Medical Sciences, especially Kurt McBurney for his feedback. This book was also inspired by my students at the University of Victoria with special acknowledgements to the UVic BioDev student club and the UVic Formula Motorsport team that I advise.

I am always grateful to my former advisors from my graduate and post-doctoral training, including Shelly Sakiyama-Elbert, Adam Arkin, and David Schaffer, who helped shape me into the researcher that I am today.

In terms of inspiration outside of science, my nephews, Tommy Willerth, Leo, and Boston Lukosky, are always amusing along with my beloved Kansas City Royals and the United States Women's National Soccer team. Special thanks to Formula One and NASCAR along with my favorite drivers Jenson Button and Brad Keselowski for keeping me amused during the lengthy writing process. I would also like to thank my housemates during my time in Madison—Glenn Trudel and his cat Lilly—for providing a creative environment.

I would also like to thank some friends from outside of academic circles, including Doug Argatoff, David Chan, John Dedman, John Douglas, André Dürkop, Brandon Edwards, Rick Ganan, Paul Jagman, David Kaempfer, Dan Kosik, John Kimball, John Olsakovsky, Duncan Phillips, Rob Stockman, Jeff Whitmore, and Red.

Last but not the least, I would like to thank my husband, Gary Escudero, for his constant support in putting up with me and my academic tendencies since 1999.

CHAPTER

The need for engineering neural tissue using stem cells 1

1 OVERVIEW OF THE NERVOUS SYSTEM

Our nervous system controls all of our actions, including those that are voluntary and those that are involuntary [1]. The complex nature of the nervous system enables us to sense and respond to our environment. Fig. 1 shows the different components that compose the nervous system, including the central nervous system and the peripheral nervous system. The central nervous system consists of the brain and spinal cord, while the peripheral nervous system contains the nerves located outside of these two organs. Several important differences exist between the central and the peripheral nervous system, which will be explored in depth in Chapter 2. In general, the central nervous system possesses a lower capacity for regeneration, while the peripheral nervous system has a higher capacity for regeneration after injury. Proper function of the nervous system requires that these tissues and organs retain their important structural features as well as the metabolic functions of the various cell types. Disruptions of these structures and their associated functions can have devastating consequences, leading to manifestation of diseases and disorders of the nervous system. This chapter will introduce some of the most common diseases and disorders that affect the nervous system along with the current treatment options and how stem cells could potentially cure these disorders.

2 DEMAND FOR CURES FOR NEURODEGENERATIVE DISEASES AND DISORDERS USING NEURAL TISSUE ENGINEERING

As the population ages, the costs associated with the healthcare required to treat patients with diseases and disorders will also continue to rise along with demand for treatments. This section will give a brief overview of the most common nervous system diseases and disorders along with the current state of treatments associated with them. Table 1 summarizes some of the major diseases and disorders that afflict the central nervous system and their associated healthcare burden. I have chosen to focus on the statistics of those suffering from these diseases and disorders in the United States and Canada due to their availability. A list of the websites for the foundations and other relevant organizations discussed in this chapter is given at the end of this chapter. Most of the treatments for these neurological diseases and disorders only alleviate the symptoms and do not represent a cure as detailed in

Engineering Neural Tissue from Stem Cells. http://dx.doi.org/10.1016/B978-0-12-811385-1.00001-7

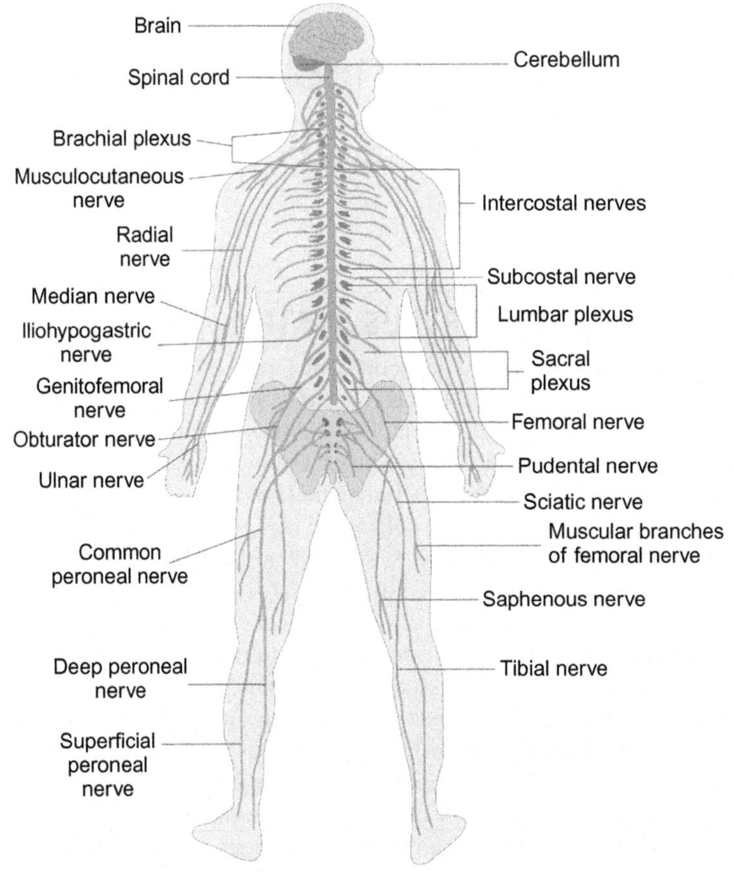

FIG. 1

Schematic of the nervous system. The brain and spinal cord comprise the central nervous system while the other nerves comprise the peripheral nervous system.

Figure produced by The Emirr. This file is licensed under the Creative Commons Attribution 3.0 Unported license.

the following sections. This book will explore how stem cells could potentially be used to develop cures for such complex diseases and disorders. Alzheimer's disease remains the most common neurological disease in North America with other major diseases of the central nervous system being Parkinson's disease, Huntington's disease, amyotrophic lateral sclerosis (ALS), and multiple sclerosis. A significant healthcare burden also exists due to injuries to the central nervous system, such as traumatic brain injury and spinal cord injury. Fig. 2 illustrates the relative incidence of each of these diseases and disorders in the form of a word cloud for easy visualization. The font size of each disease and disorder listed is proportional to the number of people suffering from that particular issue.

Table 1 List of the Most Common Neurological Diseases and Disorders Along With Relevant Statistics

Disease/Disorder	Number of People Suffering	Total Healthcare Burden	Source
Alzheimer's disease	~5,000,000 (USA)	~$236,000,000,000	Alzheimer's Association
	~564,000 (Canada)	~$10,400,000,000	Alzheimer Society Canada
Parkinson's disease	~1,500,0000 (USA)	~$25,000,000,000	Parkinson's Disease Foundation
	~100,000 (Canada)	~$558,100,000	Parkinson Canada
Huntington's disease	~30,000 (USA)	~$600,000,000[a]	Huntington's Disease Society of America
	~5000 (Canada)	~$100,000,000[a]	[2]
Amyotrophic lateral sclerosis	~20,000 (USA)	~$256–43,300,000	ALS Association/Muscular Dystrophy Association
	~3000 (Canada)	~$150,000,000	ALS Canada
Multiple sclerosis	~350,000–500,000 (USA)[b]	Not available	National MS Society
	~100,000 (Canada)	Not available	MS Society Canada
Traumatic brain injury	~1,700,000 annually (USA)	~$60,000,000,000	Center for Disease Control
	~1,000,000 (Canada)	Not available	Brain Injury Canada
Spinal cord injury	~282,000 (USA)	~$60,000,000,000	National Spinal Cord Injury Statistical Centre
	~86,000 (Canada)	~$2,700,000,000	Rick Hansen Foundation

[a] *Based on an average cost per patient of $20,000.*
[b] *The National MS Society is currently performing a study that will more accurately determine the prevalence of this disease.*

Alzheimer'sDisease
Parkinson'sDisease
AmyotrophicLateralSclerosis
Huntington'sDisease
TraumaticBrainInjury
MultipleSclerosis
SpinalCordInjury

FIG. 2

Word Cloud showing the relative incidences of the diseases and disorders of the central nervous system based on font size. Alzheimer's disease is the most common nervous system disorder in terms of patient numbers.

The diseases and disorders of the nervous system have generated significant interest in finding cures from both government and private organizations. For example, President Obama launched the White House BRAIN Initiative (Brain Research through Advancing Innovative Neurotechnologies) in 2013. The overall goal of this 12-year initiative is to develop the tools and technologies for understanding how the brain works. This effort has led to comparisons with the Human Genome Project, which greatly advanced our understanding of genetics and biology by sequencing the human genome. In Canada, private organizations, such as the Brain Canada Foundation and the W. Garfield Weston Foundation, support research in understanding how the brain works and how these different neurological diseases could be treated. These diseases and disorders of the nervous system are often complicated in terms of their biology, and they will require significant interventions to cure as detailed in the following sections.

2.1 ALZHEIMER'S DISEASE

Alzheimer's disease represents one of the most common neurological diseases in North America. Approximately one out of every nine Americans over the age of 65 suffers from this disease with the associated healthcare costs estimated to be $246 billion each year for the United States and Canada as reported by the Alzheimer's Association and the Alzheimer Society Canada. Alzheimer's disrupts the normal connections between brain cells, resulting in dementia. It is characterized by the presence of plaques and tangles in the tissue of the brain along with the degradation of cells known as basal forebrain cholinergic neurons [3]. While a positive diagnosis can currently only be confirmed postmortem based on tissue analysis, patients can be given cognitive tests to assess their mental state in terms of dementia, which occurs when normal brain function is disrupted [4]. More recently, scientists have been investigating different modalities for imaging brain tissue to detect the presence of the plaques and tangles associated with Alzheimer's. While the disease is not directly fatal, the effects of dementia lead to other physical issues, such as an inability to balance and move, which in turn, lead to fatal complications. The current

Food and Drug Administration approved treatments for Alzheimer's disease include five pharmacological drugs that regulate brain chemistry to restore proper communication in the brain that was lost due to dementia. These treatments include four types of cholinesterase inhibitors and one type of NMDA (*N*-methyl-D-aspartate) receptor antagonist [5]. Both drugs modulate the levels of neurotransmitters, chemicals released by nerve cells that enable communication with other nerve cells. These medications have varying degrees of effectiveness depending on the patient, and they only treat the symptoms of this disease. They do not treat the underlying neuronal degeneration that occurs, making these treatments short-term ways of dealing with this disease. Stem cells could potentially offer a long-term cure for this disease although much work remains before such therapies could be implemented [6,7]. For example, a certain subset of stem cells known as mesenchymal stem cells modulate the immune system of the body by eliminating foreign agents. These stem cells can potentially clear the plaques and tangles associated with Alzheimer's disease as a way to eliminate their effects on the nervous system and restore proper function [8].

2.2 PARKINSON'S DISEASE

Parkinson's disease causes degeneration of the central nervous system, resulting in impaired motor skills followed by cognitive defects as the disease progresses. The Canadian actor Michael Fox suffers from this disease and has established the Michael J. Fox Foundation for Parkinson's Research that supports efforts to find a cure for this disease. The symptoms of Parkinson's disease include tremors, bradykinesia, muscle rigidity, as well as impairing speech, writing, and posture [9]. These effects result from the death of the dopaminergic (DA) neurons in the brain that are responsible for controlling movement [10,11]. Similar to Alzheimer's disease, Parkinson's disease itself does not directly cause death, but instead its symptoms can lead to fatal complications [12]. Men are more likely to be diagnosed than women with Parkinson's. While most people are diagnosed with Parkinson's after the age of 50, a subset of patients suffer from early onset Parkinson's [13].

One of the most common therapeutic approaches replaces the dopamine that would normally be secreted by DA neurons, using a precursor molecule called L-DOPA [12]. Continued treatment with L-DOPA leads to dyskinesias (distorted movements), and thus long treatment options for Parkinson's disease must be investigated [14]. Researchers have investigated ways of promoting survival of DA neurons through the delivery of neurotrophic factors as a means of developing an effective strategy for long-term treatment of Parkinson's disease [15]. Specifically, glial-derived neurotrophic factor (GDNF) promotes the survival of DA neurons in animal models of Parkinson's disease as well as the differentiation of embryonic DA neurons grown in culture [16–18]. An ongoing clinical trial by Medgenesis Therapeutix delivers GDNF into the brains of patients suffering from Parkinson's using a novel convection–based technology. Once in the brain, GDNF promotes survival of DA neurons, reducing the effects of Parkinson's disease on motor impairment [19]. Proper delivery of appropriate concentrations of biologically active

GDNF into the correct region of the brain is necessary for achieving functional improvement in motor skills [20]. However, accomplishing this goal remains a challenge due to various issues, including the presence of the blood-brain barrier and the loss of GDNF due to diffusion into surrounding tissue. Stem cell therapy also could be used to replace the neurons that have died due to this disease. A clinical trial implanting fetal-derived neural tissue into the patients suffering from Parkinson's disease was conducted in the late 1980s with mixed results [21]. Current efforts are focused on translating stem cell-derived therapies, including the generation of replacement DA neurons from pluripotent stem cells, into the clinic as soon as 2017 [22].

2.3 HUNTINGTON'S DISEASE

The exact function of the huntingtin protein remains unknown. However, what is known is that a specific mutation in the gene encoding this protein leads to the development of Huntington's disease, making it a genetic disorder [23]. The huntingtin protein is quite large and the mutated version contains an additional 36 glutamine residues, making it even larger. This autosomal dominant mutation affects nerve cells leading to their death in the striatum. The symptoms of Huntington's disease include emotional issues such as depression and anxiety, loss of cognitive function, and physical symptoms, like loss of coordination, which are similar to those experienced by people suffering from Parkinson's disease [24]. Symptoms get progressively worse over time with most patients dying 15–25 years after the disease begins to manifest. The average age at onset is 40 years old. Current treatments only focus on treating the symptoms—the main drug prescribed for Huntington's disease is tetrabenazine, a dopamine-depleting agent that treats the physical symptoms of the disease [24]. Some groups have explored cell therapy as a way to replace the dying nerve cells associated with Huntington's disease [25]. In particular, the California Institute for Regenerative Medicine is currently funding a number of projects that develop different cell therapies to treat this disease. More information on these studies can be found at their fact sheet (https://www.cirm.ca.gov/our-progress/disease-information/huntingtons-disease-fact-sheet). Interestingly, stem cells derived from patients suffering from Huntington's disease serve as a tool to study the basic biology of this disease, which could provide greater insight into how to find successfully a long-term cure for this disease [26]. Another method for potentially curing a genetic disorder like Huntington's disease is to use gene-editing tools to correct the defective gene in the genome and thus prevent the disease from manifesting [27]. Such gene-editing tools will be discussed in Chapter 8.

2.4 AMYOTROPHIC LATERAL SCLEROSIS

ALS, also known as Lou Gehrig's disease, causes rapid, progressive degeneration of the nervous system [28]. This disease gained significant attention worldwide in 2015, as people all over the world performed ice bucket challenges shared on social media to raise awareness of ALS [29]. The term sclerosis refers to a hardening of tissues with the other terms referring to the location of the muscles affected by this disease. Specifically, cells found in the brain and spinal cord known as motor neurons begin to die off. These cells control our muscles and their death results in weakened muscles that eventually leads to death. The disease progresses rapidly with an average survival time of 3 years and no effective treatments exist currently. While the disease can strike at any time, average age at onset is in the late 40s, with environmental factors, such as serving in the military, increasing the risk of catching this disease [30].

The majority of the reported cases of ALS have no genetic basis with about 10% occurring due to familial inheritance caused by mutations in genes like superoxide dismutase 1 (SOD1) [31]. While it might be possible to use gene therapy to correct these mutations, it still only represents a small fraction of the total patient population. Currently, the only available treatment for ALS consists of the drug Riluzole, which improves the patient's ability to breathe and can extend their lifespan by 2–3 months [32]. However, it is expensive and its efficacy has been debated. In addition, attempts to develop more effective drugs for treating this disorder have not been successful [33]. Thus, many groups have explored the use of stem cell therapies as a way to replace the motor neurons lost to ALS [34]. Other research groups use patient-derived stem cell lines to screen potential drug candidates for efficacy as well [31]. Thus, further development of these stem cell technologies provides potential avenues for finding a long-term cure for ALS.

2.5 MULTIPLE SCLEROSIS

A combination of genetic and environmental factors cause multiple sclerosis, an autoimmune disorder where the body's immune system attacks the myelin sheaths that surround the nerves [35]. Myelin, a protein expressed in the cells that protect and insulate nerves, contributes to the fatty region of the nervous system referred to as white matter [36]. The typical age of onset is around the age of 34, but it can afflict individuals from a large range of ages. This disease can be difficult to diagnose as the symptoms can appear intermittently. These symptoms include fatigue, numbness, issues with vison, and bladder problems [37]. Canada has the highest rate of people suffering from multiple sclerosis in the world, suggesting that its environment plays a strong component in triggering the onset of multiple sclerosis. The most accurate way to confirm if a person has multiple sclerosis is to use magnetic resonance imaging (MRI) to detect lesions present in the nervous system [38]. Similar to ALS, no effective treatment exists for curing multiple sclerosis. The currently available treatment consists of intravenous infusions of the monoclonal antibody natalizumab [37]. However, this treatment costs ~\$60,000 per year, creating a significant burden both financially and in terms of the delivery of this treatment. Stem cell therapy offers hope for treating multiple sclerosis with potential therapeutic avenues such as modulating the immune system to prevent attacks on the nervous system, as well as the possibility of designing a cell therapy to replace the cells of the nervous system damaged by this disease [39].

2.6 TRAUMATIC BRAIN INJURY

In addition to the aforementioned diseases, trauma to the brain and spinal cord can also result in severe impairment of the functions of the nervous system. One of the most famous racing drivers in history, 7-time Formula One world champion Michael Schumacher, tragically suffered from such a traumatic brain injury while skiing in 2013 and remains in recovery to this day. Additional awareness around the effects of traumatic brain injury has been raised due to players suffering these injuries in the National Football League and in the National Hockey League [40,41]. Brain injuries occur on a spectrum, ranging from having mild impact on cognitive function to causing death. In North America, traumatic brain injuries account for 30% of all injury-related deaths according to the Centre for Disease Control. These injuries can result in a variety of symptoms, including cognitive impairment, depression, anxiety, and other mood disorders [42]. Physiologically, lesions can vary in size and may or may not be detectable using imaging technologies depending on the nature of the injury. These injuries can result in blood hemorrhaging and injury to the cells of the nervous system, which results in presentation of these symptoms associated with such injuries. Often after the initial injury, a second wave of effects and damage occurs due to the immune system responding to the lesion [43]. The blood–brain barrier, which separates the central nervous system from the rest of the body, can become disrupted and inflammation can occur [44]. A wide variety of treatments have been explored for treating these issues due to the complex nature of traumatic brain injuries. Multiple avenues for treatment exist including the use of pharmaceuticals to treat inflammation, promote regeneration, or treat the behavioral issues that result from such injuries [44]. In addition, tissue engineering approaches using stem cells could potentially be used to replace regions of damage, which could be used in combination with therapy to deal with the cognitive issues associated with such injuries [45]. More on how tissue engineering using stem cells can be used to generate replacements for damaged neural tissue will be discussed in Section 4 and in Chapters 5–7.

2.7 TRAUMATIC SPINAL CORD INJURY

The spinal cord serves as the main conduit for relaying information from the brain to the extremities, playing critical roles in the function of a healthy nervous system. Traumatic injury to the spinal cord can have devastating effects, including paralysis [46]. Similar to traumatic brain injury, the initial trauma causes death of the cells present in the spinal cord followed by secondary injury from the resulting inflammation [47]. In North America, several major research centers are dedicated to spinal cord injury research, including the Miami Project to Cure Paralysis, the International Collaboration on Repair Discoveries located in British Columbia, and the Kentucky Spinal Cord Injury Research/Spinal Cord and Brain Injury Research Center. In addition, several foundations have been established to fund research for discovering treatments for spinal cord injury including the Rick Hansen Foundation, the Christopher Reeve Foundation, and Wings for Life. More information on these organizations can be found at the end of the chapter.

The loss of cells from the initial injury can lead to the formation of a cystic cavity at the injury site that generates a physical barrier to spontaneous regeneration [47]. Astrocytes form a glial scar around this cystic cavity, isolating the injury site, and these cells begin to express inhibitory molecules [48]. These molecules inhibit other cells from migrating into the injury site, preventing regeneration attempts by the spared tissue. In addition to the presence of the glial scar, other inhibitory molecules present after such injury include myelin-associated inhibitory molecules released by damaged oligodendrocytes, cytotoxic neurotransmitters released by damaged neurons, and cytokines present as a part of the inflammatory response to injury [49].

Current clinical approaches to treating spinal cord injury include managing the injury through surgery to stabilize the spinal column while decompressing the spinal cord, and to have patients undergo physical rehabilitation [50]. Steroids, such as methylprednisolone, can be administered to reduce inflammation, but their use remains controversial [51]. Thus, current treatment options for traumatic spinal cord injury are quite limited and more advanced treatment options must be investigated to restore function lost after such injuries. The use of stem cell therapies for the treatment of SCI has demonstrated some promise in clinical trials [52,53], suggesting the potential for a long-term cure for these injuries. However, the cost of such transplants remains high, and thus new technologies for stem cell culture and expansion are necessary before such therapies can become the standard of care for such injuries.

3 THE DISCOVERY OF STEM CELLS AND THEIR IMPACT ON TISSUE ENGINEERING

The concept of stem cell revolves around a cell possessing two distinct properties—being able to produce more stem cells and the ability to generate multiple cell types [54]. An interdisciplinary team of Canadian scientists confirmed the existence of hematopoietic stem cells in 1960s in a classic set of papers [55–58]. These stem cells are involved in the maintenance of our immune system and their ability to replicate enables our body to constantly fight off disease and foreign infections. This team consisted of the Canadian duo, the biophysicist Dr. James Till and the cellular biologist Dr. Ernst McCulloch when they were working as scientists at the Ontario Cancer Institute. There was increased interest at the time in discovering the relationship between radiation and cancer due to the threat of nuclear war and its effects on populations. Their work will be discussed in further detail in Chapter 3. Soon after their discovery, Joseph Altman and Gopal Das demonstrated the presence of neural stem cells in the brain of rats [59,60]. Dr. Brent Reynolds and Dr. Samuel Weiss at the University of Calgary later confirmed the presence of neural stem cells in the human brain [61]. In parallel to these efforts to identify tissue-specific stem cells, scientists were also discovering the existence of pluripotent stem cells, which have a unique property.

Unlike hematopoietic stem cells and neural stem cells which only give rise to a limited number of cell types, pluripotent stem cells can give rise to any type of cell found in an organism. This property is known as pluripotency. In 1981, the first pluripotent stem cells termed embryonic stem cells were isolated from the inner cell mass of a mouse blastocyst [62]. These cells could give rise to all three germs layers present during development. Human embryonic stem cells were derived by the Thomson group at the University of Wisconsin-Madison in 1998 [63] even though these cells were more controversial than tissue-specific stem cells as their production required the use of human embryos. The Thomson group continues to conduct innovative stem cell research.

More recently, a group of scientists demonstrated that adult cells could be reprogrammed back into a pluripotent state. These cells have been termed induced pluripotent stem cells, and their discovery contributed to Dr. Shinya Yamanaka winning the Nobel Prize in Medicine in 2012. The ascent of induced pluripotent stem cells has been astonishing with the original work being presented at the International Society for Stem Cell Research Annual Meeting in 2006 followed by the landmark paper demonstrating that it was possible to reprogram mouse fibroblasts into an embryonic stem cell-like state later that year [64]. The derivation of human induced pluripotent stem cell lines followed rapidly in the next year [65,66]. These cells are exciting for multiple reasons. First, it is now possible to derive such cells lines from patients suffering from diseases, including the neurodegenerative diseases discussed earlier in this chapter. It was now also possible to personalize these pluripotent stem cell lines by generating a stem cell line from a patient's own cells. Both possibilities have been realized now. For example, neurons derived from human-induced pluripotent stem cells have been used to screen potential drugs for the treatment of Parkinson's disease [67]. Scientists in Japan used a patient's own reprogrammed cells to treat her age-related macular degeneration with promising results [68], although this clinical trial has moved away from using the patient's own pluripotent cell line. Outside of these examples, clinical trials are ongoing that use neural stem cells, neural cells derived from human embryonic stem cells, and hematopoietic stem cells to treat various neurological diseases and disorders, making this book quite timely. This area of research full of exciting prospects for finding effective long-term treatments for the diseases and disorders discussed in this overview.

4 THE NEED FOR INTERDISCIPLINARY APPROACHES FOR TREATING NEUROLOGICAL DISORDERS

One of my primary motivations for writing this book was to approach this set of topics in a general and accessible fashion. My career has focused on working between disciplines starting from my undergraduate education where I double majored in biology and chemical engineering and then moved into the field of tissue engineering. Tissue engineering combines cells, biomaterials, and controlled drug delivery to generate tissues to replace the ones that have ceased to function properly [69]. While tissue engineering as a field is still growing, this field will help to realize fully the potential offered by stem cells by creating a supportive environment to ensure these cells perform their proper functions. The complex nature of the nervous system and stem cells requires a multifaceted approach to truly understand how we can find cures for these diseases and disorders similar to how Till and McCulloch approached identifying stem cells over 50 years ago. As detailed in the previous section, stem cells have great potential for several of these disorders and clinical trials to evaluate the effectiveness of stem cell therapies are ongoing and being planned. Solving such complex problems will require interdisciplinary teams consisting of scholars from a wide variety of backgrounds. In addition, the advent of new gene-editing technology like CRISPR and the wide spread adoption of optogenetics, a method for controlling cell behavior using light, has given us new tools to better understand how the nervous system works—both when it is healthy and when it is diseased. This book is intended to serve as a gateway to this exciting field and additional resources will be given at the end of each chapter for further reading. Each chapter serves an overview and introduction to diverse set of topics, presented in an easy to understand format. Chapter 2 provides an introduction to the structure, function, and cells that make up the nervous system, while Chapter 3 covers the relevant stem cell types for applications in neural tissue engineering. Chapter 4 discusses design considerations when engineering neural tissue from stem cells. Then in Chapters 5 and 6, relevant examples of engineering neural tissue using stem cells are discussed in depth. Chapter 5 focuses on the use of naturally derived biomaterials for such applications, while Chapter 6 focuses on synthetic biomaterials. Chapter 7 provides analysis of different types of drug delivery systems and how they can be applied to engineering neural tissue from stem cells. Finally, in Chapter 8, new technologies, including organoid formation, 3D printing, direct reprogramming, optogentics, and gene editing, are reviewed as a way to gain insight into how these technologies will enable the next generation of strategies for engineering neural tissue. I hope that you find this field of research as exciting as I do!

RELEVANT WEBSITES

ALS Association (USA): http://www.alsa.org/
ALS Canada: http://www.als.ca/
Alzheimer's Association (USA): http://www.alz.org/
Alzheimer Society Canada: http://www.alzheimer.ca/
BRAIN Initiative: https://www.whitehouse.gov/BRAIN
Brain Injury Canada: http://braininjurycanada.ca/
Center for Disease Control: http://www.cdc.gov/
Christopher Reeve Foundation: https://www.christopherreeve.org/
Huntington's Disease Society of America: http://hdsa.org/
Huntington Society of Canada: http://www.huntingtonsociety.ca/
Kentucky Spinal Cord Injury Research: http://louisville.edu/kscirc
Michael J. Fox Foundation for Parkinson's Research: https://www.michaeljfox.org/
Multiple Sclerosis Society of Canada: https://mssociety.ca/

National Multiple Sclerosis Society: http://www.nationalmssociety.org/
National Parkinson Foundation (recently merged with the Parkinson's Disease Foundation in August 2016): http://www.parkinson.org/
Parkinson Canada: http://www.parkinson.ca/
Rick Hansen Foundation: http://www.rickhansen.com/
Spinal Cord and Brain Injury Research Center: http://scobirc.med.uky.edu/
The Miami Project to Cure Paralysis: http://www.themiamiproject.org/
Wings for Life: http://www.wingsforlife.com/en/

REFERENCES

[1] Kandel ER, Schwartz JH, Jessell TM, Siegelbaum SA, Hudspeth AJ. Principles of neural science. 5th ed. New York City: McGraw-Hall; 2012.
[2] Fisher ER, Hayden MR. Multisource ascertainment of Huntington disease in Canada: prevalence and population at risk. Mov Disord 2014;29(1):105–14.
[3] Cipriani G, Dolciotti C, Picchi L, Bonuccelli U. Alzheimer and his disease: a brief history. Neurol Sci 2011;32(2):275–9.
[4] Koric L, Guedj E, Habert M, Semah F, Branger P, Payoux P, et al. Molecular imaging in the diagnosis of Alzheimer's disease and related disorders. Rev Neurol 2016;172(12):725–34.
[5] Tayeb HO, Yang HD, Price BH, Tarazi FI. Pharmacotherapies for Alzheimer's disease: beyond cholinesterase inhibitors. Pharmacol Ther 2012;134(1):8–25.
[6] Tong LM, Fong H, Huang Y. Stem cell therapy for Alzheimer's disease and related disorders: current status and future perspectives. Exp Mol Med 2015;47(3):e151.
[7] Yue C, Jing N. The promise of stem cells in the therapy of Alzheimer's disease. Translat Neurodegener 2015;4(1):8.
[8] Turgeman G. The therapeutic potential of mesenchymal stem cells in Alzheimer's disease: converging mechanisms. Neural Regen Res 2015;10(5):698.
[9] Jankovic J. Parkinson's disease: clinical features and diagnosis. J Neurol Neurosurg Psychiatry 2008;79(4):368–76.
[10] Jung YW, Hysolli E, Kim KY, Tanaka Y, Park IH. Human induced pluripotent stem cells and neurodegenerative disease: prospects for novel therapies. Curr Opin Neurol 2012;25(2):125–30.
[11] Rodriguez M, Morales I, Rodriguez-Sabate C, Sanchez A, Castro R, Brito JM, et al. The degeneration and replacement of dopamine cells in Parkinson's disease: the role of aging. Front Neuroanat 2014;8:80.
[12] Stacy M, Jankovic J. Current approaches in the treatment of Parkinson's disease. Annu Rev Med 1993;44:431–40.
[13] Koerts J, König M, Tucha L, Tucha O. Working capacity of patients with Parkinson's disease—a systematic review. Parkinsonism Relat Disord 2016;27:9–24.
[14] Finlay CJ, Duty S, Vernon AC. Brain morphometry and the neurobiology of levodopa-induced dyskinesias: current knowledge and future potential for translational pre-clinical neuroimaging studies. Front Neurol 2014;5:95.
[15] Ungerstedt U, Arbuthnott GW. Quantitative recording of rotational behavior in rats after 6-hydroxy-dopamine lesions of the nigrostriatal dopamine system. Brain Res 1970;24(3):485–93.

[16] Lin LFH, Doherty DH, Lile JD, Bektesh S, Collins F. Gdnf—a glial-cell line derived neurotrophic factor for midbrain dopaminergic-neurons. Science 1993;260(5111):1130–2.
[17] Beck KD, Valverde J, Alexi T, Poulsen K, Moffat B, Vandlen RA, et al. Mesencephalic dopaminergic-neurons protected by Gdnf from Axotomy-induced degeneration in the adult brain. Nature 1995;373(6512):339–41.
[18] Grondin R, Gash DM. Glial cell line-derived neurotrophic factor (GDNF): a drug candidate for the treatment of Parkinson's disease. J Neurol 1998;245:P35–42.
[19] Jankovic J, Aguilar LG. Current approaches to the treatment of Parkinson's disease. Neuropsychiatr Dis Treat 2008;4(4):743–57.
[20] Iancu R, Mohapel P, Brundin P, Paul G. Behavioral characterization of a unilateral 6-OHDA-lesion model of Parkinson's disease in mice. Behav Brain Res 2005;162(1):1–10.
[21] Freed CR, Greene PE, Breeze RE, Tsai WY, DuMouchel W, Kao R, et al. Transplantation of embryonic dopamine neurons for severe Parkinson's disease. N Engl J Med 2001;344(10):710–9.
[22] Barker RA, Parmar M, Kirkeby A, Björklund A, Thompson L, Brundin P. Are stem cell-based therapies for Parkinson's disease ready for the clinic in 2016? J Parkinson's Dis 2016;1–7 [Preprint].
[23] MacDonald ME, Ambrose CM, Duyao MP, Myers RH, Lin C, Srinidhi L, et al. A novel gene containing a trinucleotide repeat that is expanded and unstable on Huntington's disease chromosomes. Cell 1993;72(6):971–83.
[24] Frank S. Treatment of Huntington's disease. Neurotherapeutics 2014;11(1):153–60.
[25] Choi KA, Hwang I, Hs P, Oh SI, Kang S, Hong S. Stem cell therapy and cellular engineering for treatment of neuronal dysfunction in Huntington's disease. Biotechnol J 2014;9(7):882–94.
[26] Tousley A, Kegel-Gleason KB. Induced pluripotent stem cells in Huntington's disease research: progress and opportunity. J Huntington's Dis 2016;5(2):99–131.
[27] Yang W, Tu Z, Sun Q, Li X-J. CRISPR/Cas9: implications for modeling and therapy of neurodegenerative diseases. Front Mol Neurosci 2016;9:30.
[28] Tefera TW, Borges K. Metabolic dysfunctions in amyotrophic lateral sclerosis pathogenesis and potential metabolic treatments. Front Neurosci 2017;10:611.
[29] Morgan S, Orrell RW. Pathogenesis of amyotrophic lateral sclerosis. Br Med Bull 2016;119(1):87–98.
[30] Beard JD, Kamel F. Military service, deployments, and exposures in relation to amyotrophic lateral sclerosis etiology and survival. Epidemiol Rev 2014;mxu001.
[31] Picher-Martel V, Valdmanis PN, Gould PV, Julien J-P, Dupré N. From animal models to human disease: a genetic approach for personalized medicine in ALS. Acta Neuropathol Commun 2016;4(1):1.
[32] Miller RG, Mitchell J, Lyon M, Moore DH. Riluzole for amyotrophic lateral sclerosis (ALS)/motor neuron disease (MND). Cochrane Libr 2007;4(1): CD001447.
[33] DeLoach A, Cozart M, Kiaei A, Kiaei M. A retrospective review of the progress in amyotrophic lateral sclerosis drug discovery over the last decade and a look at the latest strategies. Expert Opin Drug Discovery 2015;10(10):1099–118.
[34] Haidet-Phillips AM, Maragakis NJ. Neural and glial progenitor transplantation as a neuroprotective strategy for amyotrophic lateral sclerosis (ALS). Brain Res 1628;2015:343–50.
[35] Jörg S, Grohme DA, Erzler M, Binsfeld M, Haghikia A, Müller DN, et al. Environmental factors in autoimmune diseases and their role in multiple sclerosis. Cell Mol Life Sci 2016;73:1–12.
[36] Li J, Zhang L, Chu Y, Namaka M, Deng B, Kong J, et al. Astrocytes in oligodendrocyte lineage development and white matter pathology. Front Cell Neurosci 2016;10:1–13.

[37] van Pesch V, Sindic CJ, Fernández O. Effectiveness and safety of natalizumab in real-world clinical practice: review of observational studies. Clin Neurol Neurosurg 2016;149:55–63.

[38] Bou Fakhredin R, Saade C, Kerek R, El-Jamal L, Khoury SJ, El-Merhi F. Imaging in multiple sclerosis: a new spin on lesions. J Med Imaging Radiat Oncol 2016;60(5):577–86.

[39] Meamar R, Dehghani L, Nematollahi S, Tanhaee A. The role of stem cell therapy in multiple sclerosis: an overview of the current status of the clinical studies. Int J Pediatr 2014;2(2.3):72.

[40] Lehman EJ, Hein MJ, Gersic CM. Suicide mortality among retired National Football League players who played 5 or more seasons. Am J Sports Med 2016; http://dx.doi.org/10.1177/0363546516645093.

[41] Echemendia RJ, Bruce JM, Meeuwisse W, Comper P, Aubry M, Hutchison M. Long-term reliability of ImPACT in professional ice hockey. Clin Neuropsychol 2016;30(2):311–20.

[42] ALC Z, Vicentini JE, Fregni F, Rodrigues PA, Botelho C, de Lucia MCS, et al. Updates and current perspectives of psychiatric assessments after traumatic brain injury (TBI). Front Psych 2016;7:95.

[43] McKee CA, Lukens JR. Emerging roles for the immune system in traumatic brain injury. Front Immunol 2016;7:1–17.

[44] Pearn ML, Niesman IR, Egawa J, Sawada A, Almenar-Queralt A, Shah SB, et al. Pathophysiology associated with traumatic brain injury: current treatments and potential novel therapeutics. Cell Mol Neurobiol 2016;1–15.

[45] Tian L, Prabhakaran MP, Ramakrishna S. Strategies for regeneration of components of nervous system: scaffolds, cells and biomolecules. Regen Biomater 2015;rbu017.

[46] Jain NB, Ayers GD, Peterson EN, Harris MB, Morse L, O'Connor KC, et al. Traumatic spinal cord injury in the United States, 1993-2012. JAMA 2015;313(22):2236–43.

[47] Fitch MT, Doller C, Combs CK, Landreth GE, Silver J. Cellular and molecular mechanisms of glial scarring and progressive cavitation: in vivo and in vitro analysis of inflammation-induced secondary injury after CNS trauma. J Neurosci 1999;19(19):8182–98.

[48] Anderson MA, Ao Y, Sofroniew MV. Heterogeneity of reactive astrocytes. Neurosci Lett 2014;565:23–9.

[49] Silver J, Schwab ME, Popovich PG. Central nervous system regenerative failure: role of oligodendrocytes, astrocytes, and microglia. Cold Spring Harb Perspect Biol 2015;7(3):a020602.

[50] Siddiqui AM, Khazaei M, Fehlings MG. Translating mechanisms of neuroprotection, regeneration, and repair to treatment of spinal cord injury. Prog Brain Res 2015;218:15–54.

[51] Cheung V, Hoshide R, Bansal V, Kasper E, Chen CC. Methylprednisolone in the management of spinal cord injuries: lessons from randomized, controlled trials. Surg Neurol Int 2015;6:142.

[52] Scott CT, Magnus D. Wrongful termination: lessons from the Geron clinical trial. Stem Cells Transl Med 2014;3(12):1398–401.

[53] Buzhor E, Leshansky L, Blumenthal J, Barash H, Warshawsky D, Mazor Y, et al. Cell-based therapy approaches: the hope for incurable diseases. Regen Med 2014;9(5):649–72.

[54] Willerth SM. Neural tissue engineering using embryonic and induced pluripotent stem cells. Stem Cell Res Ther 2011;2(2):17.

[55] Till JE, McCulloch EA, Siminovitch L. A stochastic model of stem cell proliferation, based on the growth of spleen colony-forming cells. Proc Natl Acad Sci 1964;51(1):29–36.

[56] Becker AJ, McCulloch EA, Till JE. Cytological demonstration of the clonal nature of spleen colonies derived from transplanted mouse marrow cells. Nature 1963;197:452–4.

[57] Till JE, McCulloch EA. A direct measurement of the radiation sensitivity of normal mouse bone marrow cells. Radiat Res 1961;14(2):213–22.

[58] McCulloch EA, Till JE. The radiation sensitivity of normal mouse bone marrow cells, determined by quantitative marrow transplantation into irradiated mice. Radiat Res 1960;13:115–25.

[59] Altman J, Das GD. Autoradiographic and histological studies of postnatal neurogenesis. I. A longitudinal investigation of the kinetics, migration and transformation of cells incoorporating tritiated thymidine in neonate rats, with special reference to postnatal neurogenesis in some brain regions. J Comp Neurol 1966;126(3):337–89.

[60] Altman J, Das GD. Autoradiographic and histological evidence of postnatal hippocampal neurogenesis in rats. J Comp Neurol 1965;124(3):319–35.

[61] Reynolds BA, Weiss S. Generation of neurons and astrocytes from isolated cells of the adult mammalian central nervous system. Science 1992;255(5052):1707.

[62] Evans MJ, Kaufman MH. Establishment in culture of pluripotential cells from mouse embryos. Nature 1981;292(5819):154–6.

[63] Thomson JA, Itskovitz-Eldor J, Shapiro SS, Waknitz MA, Swiergiel JJ, Marshall VS, et al. Embryonic stem cell lines derived from human blastocysts. Science 1998;282(5391):1145–7.

[64] Takahashi K, Yamanaka S. Induction of pluripotent stem cells from mouse embryonic and adult fibroblast cultures by defined factors. Cell 2006;126(4):663–76.

[65] Takahashi K, Tanabe K, Ohnuki M, Narita M, Ichisaka T, Tomoda K, et al. Induction of pluripotent stem cells from adult human fibroblasts by defined factors. Cell 2007;131(5):861–72.

[66] Yu J, Vodyanik MA, Smuga-Otto K, Antosiewicz-Bourget J, Frane JL, Tian S, et al. Induced pluripotent stem cell lines derived from human somatic cells. Science 2007;318(5858):1917–20.

[67] Peng J, Liu Q, Rao MS, Zeng X. Using human pluripotent stem cell-derived dopaminergic neurons to evaluate candidate Parkinson's disease therapeutic agents in MPP+ and rotenone models. J Biomol Screen 2013;18(5):522–33.

[68] Garber K. RIKEN suspends first clinical trial involving induced pluripotent stem cells. Nat Biotechnol 2015;33(9):890–1.

[69] Lanza R, Langer R, Vacanti JP. Principles of tissue engineering. Cambridge, MA: Academic Press; 2011.

CHAPTER

Introduction to the nervous system

2

1 INTRODUCTION

The nervous system enables an animal to send, receive, and process information between different regions of the body. It consists of two major components: the central nervous system and the peripheral nervous system. The brain serves as the command center for coordinating all our activities, and it is one of the two major organs that compose the central nervous system, along with the spinal cord. Brain size increased as higher forms of animals evolved. Thus, the complexity of the nervous system grew as the amount of information and the number of bodily actions being processed increased. Scientists have long been interested in understanding how our brain controls the rest of the body. It achieves this control by sending and receiving signals throughout the spinal cord and the peripheral nervous system.

The initial studies exploring the functions of the nervous system were often limited to observing behaviors after brain injuries and performing dissections postmortem. Phineas Gage, shown in Fig. 1, serves as a well-known example of a patient whose brain injury caused a significant shift in behavior [1]. In 1849, Gage suffered a horrible injury when an iron rod penetrated his head, causing significant damage to his brain. Amazingly, he survived, but his personality underwent a dramatic change. This change caused scientists to hypothesize about the regions of the brain responsible for such behavior based on the location of his injury. His case study suggested to scientists that discrete regions of the brain were responsible for processing different types of information.

Even today, scientists still use patients who exhibit interesting symptoms as case studies for understanding the function of the nervous system. Recent controversy has arisen over the Patient H.M., who began exhibiting severe amnesia after having surgery to treat his epilepsy. Dr. Suzanne Conklin of the Massachusetts Institute of Technology worked with him extensively to gain a deeper understanding of how memories were made. His case was so interesting that she even wrote a book about her experiences working with him [2]. After his death, his body was donated to science and examined for further evidence of how the mind works by examining his brain structure. However, in 2016, Luke Dittrich wrote a competing book entitled "Patient H.M." painting a different picture of this scientific work, including significant commentary on the ethics when working with patients in such case studies [3]. These case studies provided significant insight into how different regions of the brain function in the context of the nervous system.

Engineering Neural Tissue from Stem Cells. http://dx.doi.org/10.1016/B978-0-12-811385-1.00002-9

FIG. 1

Historical photo of Phineas Gage. This is a sixth plate-cased daguerreotype of Gage in the collection of Jack and Beverly Wilgus. The image has been laterally reversed to show the features correctly since daguerreotypes are mirror images.
This figure is reprinted with acknowledgement of the Warren Anatomical Museum in the Francis A. Countway Library of Medicine, the holders of the copyright for this image. This image was a gift from Jack and Beverly Wilgus.

Other scientists turned to examining tissue sections of nervous tissue postmortem to increase their understanding of how the nervous system functions at different scale—the level of the cell, which is the basic unit of life. The famous Nobel Prize winning Spanish scientist Santiago Ramon y Cajal drew elegant pictures of the cells found in the brain and other regions of the nervous system as seen in Fig. 2 based on the structures he observed in such tissue sections in the late 1800s and early 1900s. A complete collection of his drawings can be found at his estate's website listed in the section on additional resources located at the end of this chapter. These drawings provided fascinating insight into the structures that make up the nervous system. His Nobel Prize was received jointly with Camilo Golgi for their work in elucidating the cellular nature of nervous system [4]. Their work contributed to the theory that neurons were the basic physiologic unit of the nervous system, which is widely accepted today.

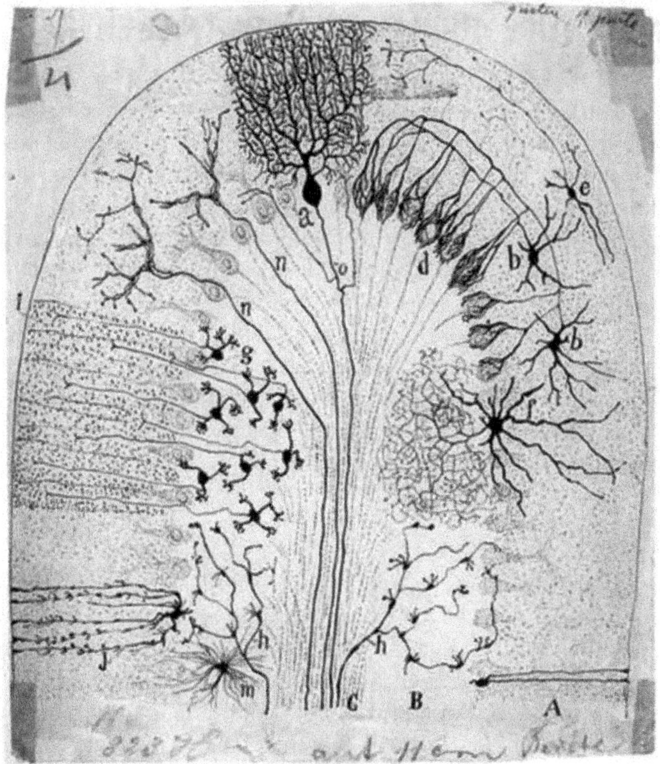

FIG. 2

Drawing by Ramon y Cajal showing a schematic cross-section of the mammalian cerebellar gyrus.

This image is reprinted with permission from the estate of Ramon y Cajal.

This chapter begins by detailing the function of different regions of the central nervous system by focusing on the brain and then the spinal cord. Next, the structures present in the peripheral nervous system will be discussed. The final section of this chapter focuses on the different types of cells found in the nervous system and how these cells enable these organs to function successfully. Understanding these cellular functions also provides a basis for developing cellular therapies for repairing the damaged nervous system. This chapter provides a general overview of neuroanatomy. For readers who desire additional detailed coverage of the nervous system, a list of resources and readings are given at the end of the chapter. Finally, a significant portion of the terminology used in this chapter comes from the traditional descriptions of human anatomy, where different parts of the body are identified using Greek or Latin to ensure the meaning is retained regardless of the language being used. The latest version of the Terminologia anatomica was published in 1998, and a link to the online version is given at the end of the chapter [5]. This reference serves as the international standard for defining the terminology used to describe human anatomy.

2 THE STRUCTURE AND FUNCTION OF THE CENTRAL NERVOUS SYSTEM

The brain and spinal cord compose the central nervous system. Each of these organs contains dedicated regions responsible for performing the specific functions associated with their activity. These functions include transmitting, receiving, and processing information from different points throughout the body. The skull and the spine serve as protective layers of bone for the brain and spinal cord respectively. These organs are delicate and essential for proper bodily function. This section begins by detailing the specific regions present in the brain followed by the covering of the relevant regions of the spinal cord. Certain terms will be used throughout this chapter and the rest of the book. For example, the terms anterior and ventral refer to the front side of the body with the terms posterior and dorsal referring to the back when discussing the anatomical features of the nervous system. These terms are often applied to the structures found in the brain and spinal cord.

2.1 BRAIN

One of the most popular myths is that we only use 10% of our brains [6]. Unfortunately, this statement is false—otherwise this section would be much shorter. While people typically think of the brain as a gray wrinkled organ, brains typically appear pink in real life due to the presence of blood vessels. This section will identify the major regions of the brain and their specific functions as shown in Fig. 3. Understanding the functions of the brain requires learning about its different subregions and their role in performing voluntary actions and maintaining proper body function in the mature nervous system. Our brain also contains specialized regions that support the maintenance of populations of stem cells. These regions will also be discussed in this section.

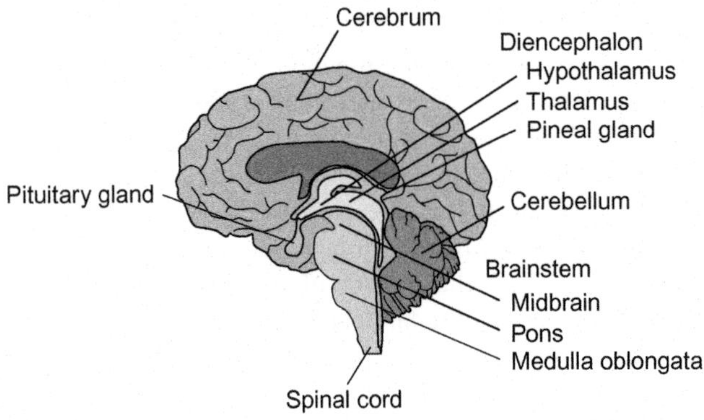

FIG. 3

Cross-section of the brain with the different regions labeled, including the cerebrum, the diencephalon, the brainstem, and cerebellum.

From Tara Styan.

The brain consists of four major regions: the cerebrum, the diencephalon, the cerebellum, and the brainstem. Accordingly, each of these regions will be discussed below under its own particular subheading. The larger regions of the brain also contain multiple zones that are further specialized to perform specific roles in the context of the nervous system. These zones will be discussed in the appropriate section below based on their location in the brain. In addition to the four regions detailed above, the brain can also be subdivided into the forebrain, midbrain, and hindbrain. These terms and their associated regions are defined based on the structures that form during the embryonic development of the brain. Additionally, the brain can be considered as being divided into two regions or lobes known as the left and right hemispheres. Failure of these different brain structures to form and separate during development can result in significant health-related issues. Such brain disorders are often referred to as encephalopathies. Understanding how the different parts of the brain work provides insight into why these encephalopathies can have such devastating effects.

2.1.1 Cerebrum

The cerebrum represents one of the largest regions of the brain as seen in Fig. 3, and its functions are critical for survival. It is responsible for processing information associated with movement, smell, sensory perception, language, communication, memory, and learning. The left and right symmetrical hemispheres present in the cerebrum are responsible for a different set of tasks. This division of labor is where the terms "left brained," meaning a person is more analytical and logical, and "right brained" where someone is more intuitive, arise—despite the lack of convincing scientific evidence to support such claims [7]. The cerebral cortex serves as the outer layer of the cerebrum and it consists of mostly of gray matter, which is a type of tissue labeled on the basis of its color [8]. Four lobes make up the cerebral cortex: the frontal lobe, the parietal lobe, the temporal lobe, and the occipital lobe. Each lobe has a distinct function. For example, the frontal lobe processes information associated with problem solving, speech, and emotions. The parietal lobe senses stimuli and movement, while the temporal lobe deals with processing auditory stimuli and speech. The occipital lobe processes visual information. Generally, this region controls voluntary action by working in coordination with the region of the brain known as the cerebellum, which is part of the brainstem.

The hippocampus, basal ganglia, and olfactory bulb are located in the deeper regions of the cerebrum, and these structures play unique roles in the brain function. The structure of the hippocampus resembles a seahorse, and accordingly it is named after the Greek word meaning seahorse. This region plays an important role in long-term memory [9]. It consists of two sections: the hippocampus proper region and the dentate gyrus. The dentate gyrus holds particular interest as it is one of the regions of the brain where adult neural stem cells are found, as well as a site of neurogenesis, which is the process of forming new neurons from stem cells [10]. This region of the brain becomes dysfunctional in patients suffering from Alzheimer's disease, and neuroscientists have been looking for connections between neurogenesis and Alzheimer's disease [11]. The basal ganglia consist of the nuclei (the command center of a cell) located laterally in a coronal section from a structure called the thalamus, which is found in the diencephalon region of the brain. These two structures work together to coordinate movement through signaling by the molecule glutamate. More on cell-to-cell signaling will be discussed in Section 4, which details the cells of the nervous system and their functions. The main functional cellular units of the nervous system are neurons. These cells rely on different types of signaling to transmit a variety of messages throughout the body. Multiple diseases and disorders are associated with improper basal ganglia function, including Parkinson's disease, attention deficit hyperactivity disorder (ADHD), and schizophrenia [12]. Some of the symptoms of these diseases manifest as disordered movement, which is consistent with the function of this region in healthy tissue.

As its name implies, the olfactory bulb plays a critical role in maintaining the sense of smell. This region contains several receptors that enable the body to sense and filter stimuli detected through olfaction [13]. This information is then transmitted to other regions of the brain where it is processed accordingly. The olfactory bulb also contains multipotent stem cells to replenish cells lost during the sensing process [14]. These neural stem cells migrate to the olfactory bulb from a region called the subventricular zone, which will be discussed later in this chapter. Interestingly, the loss of the ability to smell is observed in many neurodegenerative diseases, including dementia and Alzheimer's disease. This observation suggests a potential common link between these diseases caused by an inability of the brain to perform neurogenesis, the development of new neural tissue.

2.1.2 Diencephalon

The posterior part of the forebrain contains the diencephalon, which consists of two major structures—the thalamus and the hypothalamus. As touched on briefly, the thalamus performs the important task of processing information obtained from the rest of the nervous system and transmitting it to the cerebral cortex [15]. This egg-shaped region of the brain also regulates consciousness. The hypothalamus serves as a regulator of autonomic and endocrine functions [16,17]. Defective functioning of the hypothalamus can lead to a range of diseases, including obesity, diabetes, and Alzheimer's disease [18].

2.1.3 Brainstem

The brainstem serves as the physical structure that connects the brain to the spinal cord, containing portions of the mid and hindbrain regions. This structure plays several significant roles in maintaining proper function of the nervous system and its control over our bodies. The brainstem regulates and maintains the cardiac and respiratory systems. Additionally, it plays an important role in controlling consciousness and the process of sleeping. Injuries to the brainstem can be life threatening, as they can result in an inability to breathe involuntary, which often will lead to death. In addition to controlling these major bodily functions, the major nerve conduction pathways (often referred to as tracts) that regulate our ability to sense start in the brain, then move through the brainstem and into the spinal cord before branching out all over the body. The nerves responsible for sensing and movement in the face branch out from the brainstem, enabling us to respond to stimuli. These nerves are known as cranial nerves because they exit the skull. The medulla oblongata, a cone-shaped region of the brainstem, serves as the continuation of the spinal cord into the skull. Accordingly, it forms the lowest region of the brainstem. It also controls the autonomic activity of the heart and lungs, including involuntary activity. Thus, it is critical to have a functioning medulla oblongata for survival. Injuries to this specific region of the brainstem are often critical and life threatening as mentioned previously, an inability to breathe can lead to death. While the pons region of the brainstem is relatively small, its function is critical in several ways. It relays messages between the cerebrum and the cerebellum to ensure proper brain activity. It regulates the processes of sleeping and dreaming as it is the location in the brain where rapid eye movement (REM) sleep associated with dreaming occurs. Overall, the brainstem performs highly critical functions for maintaining proper bodily function and survival.

2.1.4 Cerebellum

The cerebellum, the region of the brain responsible for learning motor skills, coordinates muscular activity throughout the body. It is located at the back of the skull in the region known as the hindbrain and it is connected to the brainstem. Disorders of the cerebellum affect movement, and are often referred to as ataxias (derived from the Greek word meaning out of order) [19]. This region derives its name from the phrase "little brain," reflecting its important function.

2.2 SPINAL CORD

The spinal cord consists of a collection of nerve fibers located inside the spine. It connects the medulla oblongata region of the brain with the rest of the nervous system, including the peripheral nervous system. The spinal cord acts like a highway with carloads of information entering and exiting at its different segments, depending on what region of the body is sending or receiving signals. A cross-section of the spinal cord can be seen in Fig. 4. The gray region that resembles a butterfly is referred to as gray matter while the rest of the spinal cord consists of white matter. The colors arise from differences in the cellular composition in each region. These differences will be discussed in detail in Section 4. The tracts of the spinal cord transmit different types of information and are classified accordingly. The ascending tracts start in the periphery and information travels up them in the spinal cord to the brain. These afferent tracts are associated with the body's ability to sense information in its surrounding. The other major types of tracts descend from the brain through the spinal cord and peripheral nervous system. These efferent tracts control and affect movement, often being referred to as motor tracts. While this chapter will not go into detail regarding the numerous tracts of the spinal cord, it is a general rule of thumb that tracts whose name begins with the prefix "spino" are sensory and those that end in "spinal" are motor. Examples of these modifiers include the sensory tract known as the spinocerebellar tract and the corticospinal tract, which is a motor pathway.

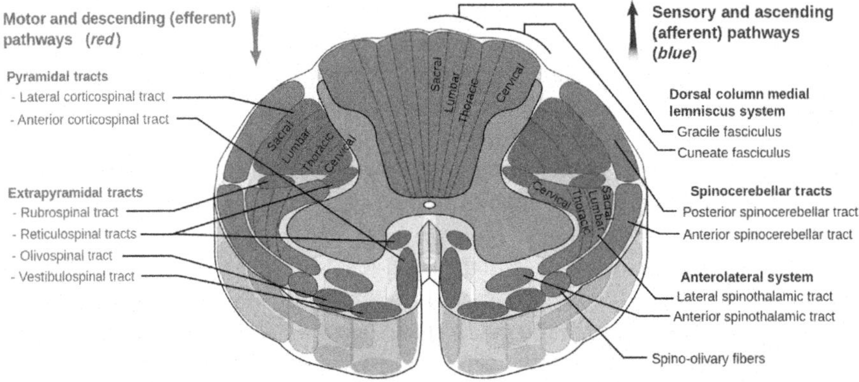

FIG. 4

Cross-section of the spinal cord showing the location of the different motor and sensory tracts present.

Figure produced by Polarlys (translation by Selket) and appears unaltered. This file is licensed under the Creative Commons Attribution 3.0 Unported license.

The three major sensory/afferent tracts include the posterior column, spinothalamic tract, and the spinocerebellar tract. These tracts enable sensing of pressure, temperature, and pain, and this information is then relayed back to the brain for processing. The two major sets of tracts associated with the motor/efferent pathways include the corticospinal tract and the extrapyramidal tracts. The corticospinal tract controls the movement of the skeletal muscles, while the extrapyramidal tracts regulate the involuntary actions required for subconscious motor control. These sensory tracts run the length of the spinal cord with each individual tract being composed of specific neuronal subtypes associated with its function. Certain tracts can also send and transmit information in both directions, which will not be discussed in this chapter.

In addition to longitudinal nerve tracts, the spinal cord can also be classified into five regions—the cervical, the thoracic, the lumbar, the sacrum, and the coccygeal—based on their position in the body as shown in Fig. 5. These five regions contain 31 individual segments with each segment containing pairs of left and right nerves. An additional 12 segments are located in the brainstem for a total of 43 segments in the entire body. These segments are covered by 33 sets of vertebrae, the regions of bone that protect the spinal cord from damage and injury. The cervical region starts at the base of the skull and runs approximately to the shoulder height. It consists of eight segments, including the ones responsible for sending and receiving information from the upper region of the body such as the arms and neck. The thoracic region has 12 segments responsible for controlling the functions in the middle of the body, including the abdomen. The lumbar region contains five segments and the nerves present in this region control leg movement and sensation. The sacrum has five segments, while the coccygeal region only has one. The sacral nerves are associated with bladder and sexual functions. As each region regulates different bodily functions, the injury

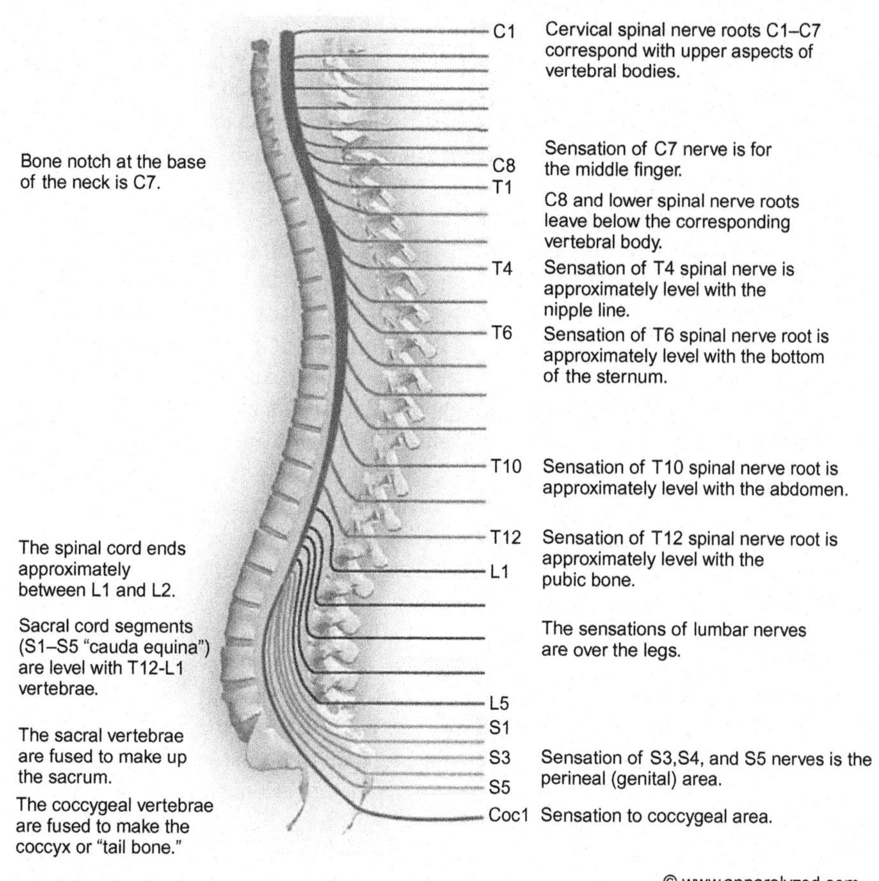

FIG. 5

The different segments of the spinal cord and their associated regions of activity.

This figure is reprinted with permission from Apparelyzed.com who retains the original rights to this image.

location during trauma to the spine plays a significant role in determining how a person will be affected. Each of these segments of the spinal cord has an associated set of roots, which is the region of nerve that leaves the spinal cord. These roots are labeled dorsal and ventral depending on their location. These spinal nerves connect the spinal cord to the peripheral nervous system, which is the focus of the next section.

3 PERIPHERAL NERVOUS SYSTEM

The peripheral nervous system consists of the somatic nervous system, sensory nervous system, and autonomic nervous system. These systems have a significantly different cellular composition compared to the central nervous system and these differences will be discussed in Section 4. The somatic nervous system controls the musculoskeletal system and our external sensory organs with this function being controlled voluntary by our brains. These actions are determined by information received from the 43 sets of sensory nerves (12 in the brainstem and 31 in the spinal cord) present in our bodies. These signals are mainly transmitted through motor neurons. In contrast, the sensory nervous system sends information obtained from external stimuli or our internal organs to the brain to be processed and interpreted. These signals are sometimes referred to as impulses and they are transmitted through sensory nerves. Understanding the different types of nerve tissues present in peripheral nervous system is essential for developing cell therapy-based strategies for repair [20]. Repairing the wrong pathways could lead to increased ability to transmit these impulses without restoring the lost function. The somatic and sensory systems can be classified together due to their association with voluntary function.

The final system of the peripheral nervous system is the autonomic nervous system, which is responsible for involuntary function [21]. As its name implies, the autonomic nervous system controls functions such as breathing, beating of the heart, digestion, and certain reflexes at local level compared to the regions of the brain responsible for these same functions. These functions are essential for the survival and accordingly they are performed at an unconscious level by the hypothalamus. These functions can be further broken down into two branches consisting of the sympathetic nervous system and the parasympathetic nervous system. The sympathetic nervous system can be broadly characterized as the tendency to "fight or flight" as observed by Walter Cannon in the early 1900s [22], while the parasympathetic nervous system was considered as regulating the need to "rest and digest." These systems have two distinct functions as the sympathetic system functions to rapidly initiate responses that are associated with spending energy and the parasympathetic systems serves to dampen such responses while minimizing the amount of energy used by an organism. Overall, the peripheral nervous system plays a critical role in carrying out the instructions sent to it from the different regions of the brain through the spinal cord. Additionally, the peripheral nervous system has a higher regenerative capacity compared to the central nervous system, which is an important consideration when developing strategies to repair damage to these systems [23].

4 CELLS OF THE NERVOUS SYSTEM

The previous sections have focused on the physical structures and their associated functions in the context of the nervous system and proper functioning of the body. The functions only occur if the cells that compose these regions and organs work together in an organized fashion. This section will address the different types of cells found in the nervous system and how they each contribute to ensure proper transmission of information throughout the body. A variety of cell types are found in both the central nervous system and the peripheral nervous system. It is important to understand these differences when developing cell therapies for the different diseases and disorders discussed in Chapter 1. These cells can be divided in two general categories: neurons and glia. The glial cells include oligodendrocytes and astrocytes in the central nervous system and Schwann cells in the peripheral nervous system. These cells play important roles in supporting the function of neurons, the major functional unit of the nervous system. The properties of these different cellular populations will be addressed in depth in the following sections.

4.1 NEURONS

It is estimated that there are 100 billion neurons located in the brain [24]. Neurons serve as the main functional unit of the nervous system and they have distinct parts that enable this property [25]. Neuronal cells possess an asymmetric quality, as they require distinct regions for accepting input and transmitting information to enable their proper function. To transmit information successfully, one neuron releases specific chemical signals referred to as neurotransmitters into a gap known as the synapse. Neurotransmitters are characterized by their ability to conduct messages in the brain, and they can be proteins, peptides, or small molecules. The neurotransmitters then cross the synapse to reach a second neuron that receives these signals by taking them up through endocytosis. The neuron sending the signal is referred to as the "presynaptic" cell, while the receiving cell is classified as being "post-synaptic." These labels are based on their locations relative to the synapse. The signals are transmitted inside the neuron in the form of action potentials, which are rapid events that occur due to a significant change in the electrical field of the cells rapidly. These action potentials enable the rapid communication of signals through the nervous system compared to other forms of biological signal transmission as they are based on electric signaling.

The structure of the neuron enables these cells to conduct these signals successfully through specialized regions of their cellular structure. As shown in Fig. 6,

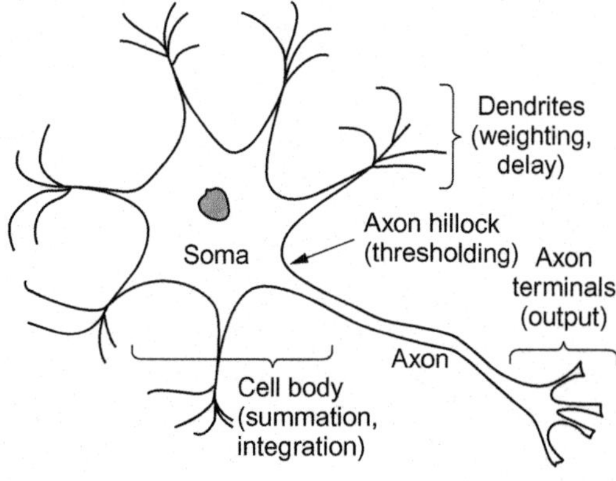

FIG. 6

Schematic showing the different parts of a neuron.

This figure is reprinted with permission of the Optical Society America. It was originally published as Fig. 1 in "Ultrafast alloptical implementation of a leaky integrate and fire neuron." Kravtsov K, Fok MP, Rosenbluth D, Prucnal PR. Opt Exp 2011;19(3):2133–2147.

these parts include the cell body (also known as the soma), the dendrite, and the axon. Neurons receive information from other neurons in the structure called the dendrite, which is the region of the cell consisting of short-branched cellular processes. The number of size of these processes determines how the neuron will obtain and process the information received. Defects in the dendrite development can lead to neurological diseases such as autism [26]. The dendrites extend from a region of the neuron referred to as the cell body. The cell body or "soma" contains the nucleus of the neuron. Certain proteins, such as neurotransmitters, must then be transported to the appropriate region of the cell where they can perform their proper function, including secretion into the synapse where they can act on other neurons. Neurons must be able to transmit signals to ensure proper functioning of the body, and disruptions in the process of transmitting these proteins can also lead to serious consequences for the nervous system [27]. These neurotransmitters traverse down the narrow region of the neuron known as the axon. The axon begins in a region known as the axon hillock, which connects the cell body to the axon. This region also serves as one of the major sites of protein synthesis in the neuron. The action potentials are generated in this region of the cell, starting once the cell has reached a certain level of stimulation from the presynaptic cell. The axon hillock also acts as a gatekeeper by determining which proteins can be transported in the axon due to the high degree of control that must be maintained over neuronal signaling to ensure proper nervous system function. The presynaptic terminal is located at the end of the axon where neurotransmitters are released. It is named based on its proximity to the synapse. Earlier in this chapter, the concepts of gray and white matter were referenced. Gray matter consists mainly of the cell bodies, dendrites, and the axon terminals associated with the neurons of the central nervous system along with unmyelinated axons. White matter obtains its color from the myelin sheaths provided by oligodendrocytes that surround the axon and the function of these sheaths is discussed in the next section.

Many different systems exist for classifying neurons based on their properties and role in the nervous system. Three broad categories of neurons exist, including sensory neurons, motor neurons, and interneurons (Fig. 7). Sensory neurons tend to have a morphology that is unipolar. These cells extend into two short processes consisting of the dendrite and axon that enable the transmission of information associated with the sensory tracts found in the nervous system. Motor neurons have a multipolar morphology as indicated by the numerous processes extended from the cell. These cells are associated with motor tracts and transmitting information related to movements. Finally, interneurons integrate the sensory and motor neurons and they also enable memory formation and decision making. These cells exhibit bipolar morphology. While sensory and motor neurons exist in both the central and peripheral nervous system, interneurons are only found in the central nervous system. Neurons can be further classified based on their role in signaling (excitatory or inhibitory) or the type of neurotransmitters they secrete. Overall, this group of cells represents an important cellular population in the nervous systems and neuronal loss often leads to the diseases and disorders discussed in Chapter 1.

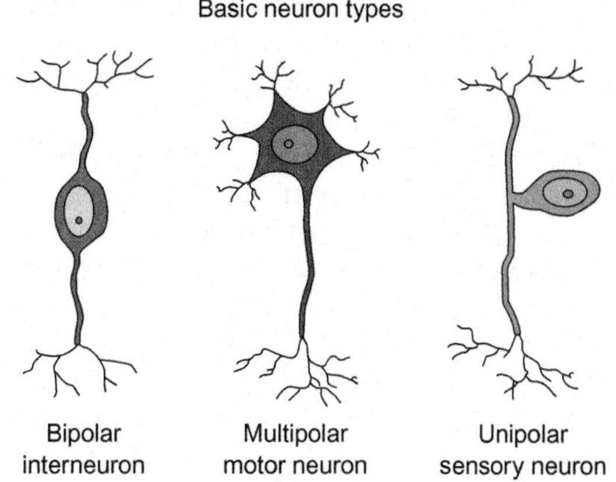

FIG. 7

Broad classifications of neurons based on structure and function.

From Tara Styan.

Thus, the production of various neuronal subtypes is often a focus of cell therapy approaches for repairing the damaged nervous system.

4.2 OLIGODENDROCYTES

Oligodendrocytes play an important role in supporting neuronal function in the central nervous system. Schwann cells play an analogous role in the peripheral nervous system, which will be discussed in Section 4.4. Their name originates from the Greek words "oligo," meaning few, and "dendron," meaning tree (Fig. 8). This description is fitting

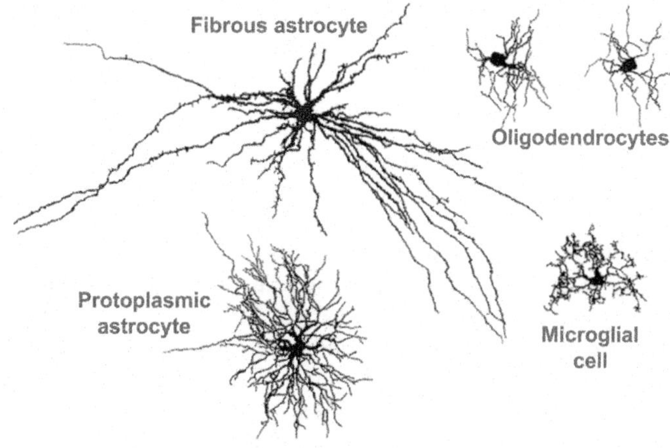

FIG. 8

Morphologies of the different types of glial cells.

Republished with permission of the original author—Thomas Fletcher.

as these cells extend few processes from their cell body, especially in comparison to neurons and astrocytes. Oligodendrocytes protect and support neurons through a variety of mechanisms. One of their most crucial roles is to generate and maintain structures known as myelin sheath [28]. The myelin sheath encompasses axons, enabling more effective transmission of their signal by acting as an insulator with a high resistance. A single oligodendrocyte can provide this support for up to 50 neurons. Gaps in the coverage of axons by the myelin sheath are referred to as nodes of Ranvier [29]. The myelin sheath consists of a mixture of proteins and lipids known as myelin. Both biomolecules are synthesized by the oligodendrocytes. In addition to providing the myelin sheath, oligodendrocytes also secrete neurotrophic factors such as brain-derived neurotrophic factor (BDNF) and glial-derived neurotrophic factor (GDNF), which further promote neuronal survival [30]. Recent studies have suggested that oligodendrocytes also exhibit heterogeneous properties based on the location in the central nervous system [31], which may have implications when developing potential cell therapies. In terms of development, oligodendrocytes are one of the last cell types found in the central nervous system to mature and are involved in the process of myelinating the brain and spinal cord [32]. In development, glial-restricted progenitors known oligodendrocyte progenitor cells (OPCs) derived from neural stem cells form during late neurogenesis. These cells can produce both oligodendrocytes and astrocytes [33,34]. Once an OPC has been directed to become a mature oligodendrocyte, it starts to extend processes while producing myelin.

While the evolution of myelin has enabled humans to have more complex nervous systems capable of processing large amounts of information, it does mean that defects and disorders in the expression and maintenance of myelin have a big impact on the physiological functioning [35]. Two major classes of myelin-related diseases include leukodystrophies and demyelinating diseases. Multiple sclerosis is one of the most prevalent diseases associated with myelin as it occurs when the immune system starts attacking oligodendrocytes [36]. Other disorders associated with myelin defects include spinal cord injury, cerebral palsy, and stroke. The use of oligodendrocyte progenitors derived from human embryonic stem cells for treating spinal cord injuries is currently being evaluated by Asterias Therapeutics [37], indicating the potential value of oligodendrocytes as a therapeutic target. Thus, understanding their function provides valuable insight into developing potential cell therapies.

4.3 ASTROCYTES

Astrocytes derive their name from the Greek words for "star" and "cell" as these cells are star shaped as seen in Fig. 8. They perform a supporting role in maintaining the proper functioning of the central nervous system [38]. These cells help to pattern the brain and provide structural support for other cells present in the central nervous system. They also provide biochemical support to neurons while ensuring proper synapse function by secreting different factors and absorbing neurotransmitters from synapses. Accordingly, they are the most common cell found in the central nervous system. Astrocytes are often associated with blood vessels and interact with the cells that make up the blood vessels through chemical signaling. In one of the major examples of this synergy, astrocytes work together with endothelial cells to generate the blood–brain barrier (BBB) and maintain its integrity [39]. The BBB serves as a selective filter, permitting only certain molecules into the central nervous system. This function enables it to act as a gatekeeper, ensuring that harmful substances cannot enter and affect the function of the central nervous system.

Astrocytes also come in different varieties based on their shape and location in the brain [38]. The three major classes of astrocytes are fibrous, protoplasmic, and radial (Fig. 8). Fibrous astrocytes are associated with white matter, and their shape is characterized by their complex network of relatively shorter processes extended from the cell body when compared to the morphology of protoplasmic astrocytes. The protoplasmic astrocytes possess fewer, longer processes and these cells are found in gray matter. During development, neural stem cells first give rise to radial glia (astrocytes) that extend to two major processes. These cells help to direct neuronal migration during neurogenesis [40]. After the developmental period, these cells transform into the aforementioned types of astrocytes. The retina contains a further specialized subset of radial astrocytes known as Müeller cells, which support the neurons present in the eye by regulating their metabolism [41].

Diseases and disorders of the central nervous system trigger the process known as astrogliosis, where astrocytes proliferate in response to the chemicals secreted by damaged neurons [42]. These astrocytes transition to a state where they are considered reactive. These reactive astrocytes then start secreting molecules that inhibit regeneration and promote scar formation. While this process is intended to preserve the function of the remaining undamaged regions of the central nervous system, its effects also prevent regeneration that might restore function from occurring. Thus, the role of reactive astrocytes and astrogliosis must be considered when determining treatments for the different diseases and disorders of the central nervous system.

In addition to astrogliosis, neurological disorders, including tauopathies and synucleinopathies, can also occur when astrocytes become dysfunctional [43]. These diseases occur when aggregates of proteins build up in the synapses, negatively affecting function by inhibiting neurotransmitter transport. This effect, in turn, reduces the functioning of the nervous system as neuronal signaling is impaired. Astrocytes and their progenitors have recently become an attractive target for cell therapy applications for treating neurological diseases and disorders now that their essential roles in the central nervous system have become better understood [44]. These therapies potentially involve transplanting healthy astrocytes to replaced diseased or reactive versions of these cells. These transplanted cells could also improve function by restoring a more permissive environment through the secretion regeneration promoting factors. Finally, it might be possible to induce adult astrogenesis directly and promote repair of the damaged nervous system without cellular transplantation [45].

4.4 SCHWANN CELLS

Schwann cells, shown in Fig. 9, play a role similar to oligodendrocytes for the neurons located in the peripheral nervous system as these cells support signal conduction

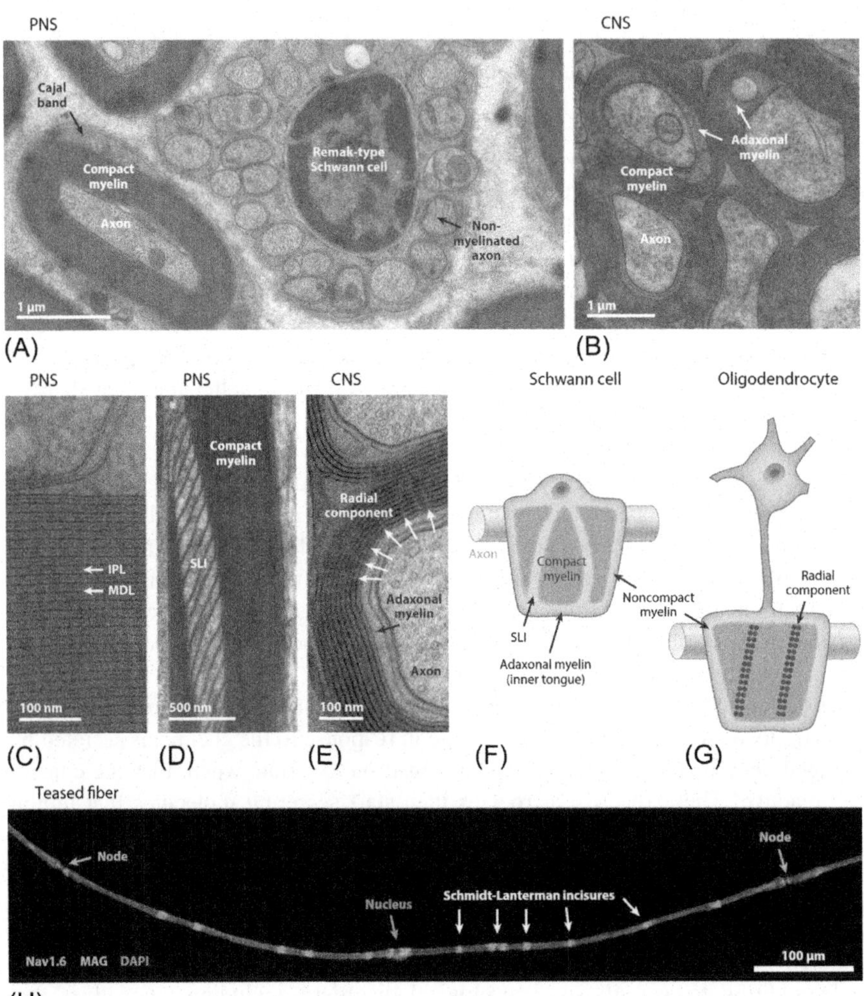

FIG. 9

Myelinated axons in the peripheral and the central nervous system (PNS and CNS, respectively). (A) Electron microscopy of the peripheral sciatic nerve shows that a myelinating Schwann cell elaborates myelin as multiple periodic membrane layers ensheathing one axonal segment. On the same image, a nonmyelinating Schwann cell engulfs multiple axons of a diameter below 1 µm. (B) In the central optic nerve, oligodendrocytes myelinate multiple axonal segments. (C) The ultrastructure of myelin is enlarged to visualize the electron-dense intraperiod lines (IPL) and major dense lines (MDL). (D) Schmidt-Lanterman incisures (SLI) provide cytosolic channels through compact peripheral myelin. (E) Radial components are adhesive tight junction strands specific to central myelin. (F and G) Peripheral and central myelin sheaths are schematically depicted as unrolled to visualize structural specializations. (H) To illustrate the dimensions of the axon/myelin unit, a single internodal segment is shown. The teased fiber was dissected from the sciatic nerve. Nodes of Ranvier were labeled with antibodies specific for the axonal sodium channel Nav1.6 *(green in the electronic version, gray in the print version)*, and Schmidt-Lanterman incisures were marked by antibodies against myelin-associated glycoprotein (MAG, *orange in the electronic version, gray in the print version*). The nucleus of the Schwann cell is labeled in *blue in the electronic version, gray in the print version.*

in axons [46]. Mature Schwann cells can be classified as myelinating or nonmyelinating. These mature cell phenotypes arise from a common immature progenitor known as a neural crest cell. Neural crest cells can both proliferate and give rise to cells found in the peripheral nervous system as they are stem cells [47]. After injury, mature Schwann cells can revert back to this proliferative, undifferentiated state (often referred to as a Büngner cell), which can promote recovery post injury [48]. Schwann cells can promote regeneration post injury as they secrete a number of neurotrophic factors, such as GDNF [49]. As with other support cells present in the nervous system, defects in the biology of Schwann cells lead to significant disorders, including Charcot-Marie-Tooth disease and Guillain-Barré syndrome [50,51]. There has been increased attention to Guillain-Barré syndrome recently as it has been linked to infection with the Zika virus [52], highlighting the intersection of the immune system with nervous system function.

On the other hand, these pro-regeneration properties of Schwann cells, including their ability to myelinate and secrete growth factors, have made them attractive for applications in cell therapy. For example, a significant amount of work has evaluated whether the transplantation of Schwann cells into the injured spinal cord promotes regeneration and clinical trials evaluating this potential are ongoing [53]. Finally, olfactory ensheathing cells are considered to be a subset of Schwann cells, despite being found in both the central and peripheral nervous system [54]. As their name suggests, these cells secrete factors that promote the formation of olfactory neurons. These cells also have been evaluated for applications in cell therapies due to these properties [55]. These applications include stroke, spinal cord injury, peripheral nerve injury, and other neurodegenerative diseases. Overall, the support cells found in the peripheral nervous system hold great potential for cell therapy applications.

4.5 OTHER CELLS OF THE NERVOUS SYSTEM

Cells called microglia exist in the brain and spinal cord to perform the role of the immune system as the central nervous system is immunoprivileged due to the protection provided by the blood-brain barrier [56]. These microglia cells are considered macrophages ("big eaters"), and they play important roles in neuronal development and maintaining homeostasis. These cells arise from the cellular populations present in bone marrow, but they function in a different manner due to their location in the central nervous system. In a healthy tissue, they constantly monitor the brain for the presence of unfolded proteins and other undesirable biomolecules and engulf them, helping to ensure proper neuronal signaling. Upon activation in response to injury, they proliferate and can perform many different functions associated with the immune response, including secreting inflammatory cytokines, recognizing foreign bodies, and taking up and presenting antigens. These cells can both simultaneously promote regeneration while causing further tissue damage due to their role in modulating inflammation and the nature of the central nervous system [57]. Modulating this process of neuroinflammation can potentially lead to better outcomes when treating diseases and disorders of the central nervous system [58].

As mentioned earlier in this chapter, two regions of the brain generate neural stem cells, which are characterized by their ability to become the mature cells of the nervous system while being able to self-replicate [10,59]. These cells play crucial roles during development and in the adult brain, and their properties and applications to cell therapy will be covered in detail in Chapter 3.

5 CONCLUSIONS

The nervous system possesses a high degree of complexity, including diverse populations of cellular phenotypes and specialized regions of function. This chapter has provided a high-level overview of the major structures found in both the central and peripheral nervous system along with the cells that contribute to its successful function. While certain cell therapy applications have been addressed briefly in this section, Chapter 3 will discuss the potential options for using stem cells to replicate the functions and structures of the healthy nervous system described here as a way to repair the damaged neural tissue.

ADDITIONAL RESOURCES

MORE READING ON HISTORY OF UNDERSTANDING NEUROANATOMY STRUCTURE AND FUNCTION

Phineas Gage background: http://brightbytes.com/phineasgage/more.html

Repository of Ramon y Cajal Drawings: http://cvc.cervantes.es/ciencia/cajal/cajal_recuerdos/recuerdos/laminas.htm

BOOKS ON PATIENT H.M.

Corkin, S., *Permanent present tense: The unforgettable life of the amnesic patient.* Vol. 1000. 2013: Basic books.

Dittrich, L., *Patient H.M.: A Story of Memory, Madness, and Family Secrets* 2016: Random House.

RESOURCES ASSOCIATED WITH GROSS NEUROANATOMY

Principles of neural science. Eds. Eric R. Kandel, James H. Schwartz, and Thomas M. Jessell. 5th edition. New York: McGraw-hill, 2012. http://www.principlesofneuralscience.com/.

GRAY'S ANATOMY (CURRENT AND ORIGINAL EDITIONS)

Standring, S. Gray's Anatomy, The Anatomical Basis of Clinical Practice. 41st edition. Elsevier, 2015. https://elsevier.ca/product.jsp?isbn=9780702052309

Gray, H. Anatomy: Descriptive and Surgical. London: John W. Parker and Sons, 1858.

Original version available through the National Institutes of Health as it is now public domain: https://collections.nlm.nih.gov/bookviewer?PID=nlm:nlmuid-06220300R-bk

Anatomy of the Brain at the University of British Columbia's Medical School: http://www.neuroanatomy.ca/MRIs/mri_coronal.html?id=1

Online version of the Terminologia anatomica: http://www.unifr.ch/ifaa/Public/EntryPage/HomePublic.html

The Hippocampus as a Cognitive Map: http://www.cognitivemap.net/

Society for Neuroscience's blog about the brain and its functions: http://blog.brainfacts.org/

RESOURCES ASSOCIATED WITH CELLULAR LEVEL NEUROANATOMY

Lodish, H., Berk, A., Kaiser, C., Krieger, M., Bretscher, A., Ploegh, H., Amon, A., and Martin, K. Molecular Cell Biology, 8th Edition. New York: W.H. Freedman. 2016.

Alberts, B., Johnson, A., Lewis, J., Morgan, D., Raff, M., Roberts, K. Walter, P. Molecular Biology of the Cell. 6th edition. New York: Garland Science. 2014. http://garlandscience.com/product/isbn/9780815344322

Alberts, B., Bray, D., Hopkin, K., Johnson, A., Lewis, J., Raff, M., Roberts, K., Walter, P. Essential Cell Biology. 4th edition. New York: Garland Science. 2013. http://www.garlandscience.com/product/isbn/9780815344544

REFERENCES

[1] Macmillan M. Restoring phineas gage: a 150th retrospective. J Hist Neurosci 2000;9(1):46–66.

[2] Corkin S. Permanent present tense: the unforgettable life of the amnesic patient. New York, NY: Basic Books; 2013.

[3] Dittrich L, Patient HM. A story of memory, madness, and family secrets. New York, NY: Random House; 2016.

[4] Fishman RS. The nobel prize of 1906. Arch Ophthalmol 2007;125(5):690–4.

[5] Terminology FCoA. Terminologia anatomica. Stuttgart: Georg Thieme Verlag; 1998.

[6] Sala SD. Mind myths: exploring popular assumptions about the mind and brain. Hoboken, NJ: John Wiley & Sons; 1999.

[7] Nielsen JA, Zielinski BA, Ferguson MA, Lainhart JE, Anderson JS. An evaluation of the left-brain vs. right-brain hypothesis with resting state functional connectivity magnetic resonance imaging. PLoS One 2013;8(8):e71275.

[8] Miller EK, Cohen JD. An integrative theory of prefrontal cortex function. Annu Rev Neurosci 2001;24(1):167–202.

[9] O'keefe J, Nadel L. The hippocampus as a cognitive map. Oxford: Oxford University Press; 1978.

[10] Gage FH. Mammalian neural stem cells. Science 2000;287(5457):1433–8.
[11] Hollands C, Bartolotti N, Lazarov O. Alzheimer's disease and hippocampal adult neurogenesis; exploring shared mechanisms. Front Neurosci 2016;10:178.
[12] Yamada K, Takahashi S, Karube F, Fujiyama F, Kobayashi K, Nishi A, et al. Neuronal circuits and physiological roles of the basal ganglia in terms of transmitters, receptors and related disorders. J Physiol Sci 2016;66:1–12.
[13] Nagayama S, Homma R, Imamura F. Neuronal organization of olfactory bulb circuits. Front Neural Circuits 2015;52:1–19.
[14] Gallarda B, Lledo P-M. Adult neurogenesis in the olfactory system and neurodegenerative disease. Curr Mol Med 2012;12(10):1253–60.
[15] Sherman SM. Thalamus plays a central role in ongoing cortical functioning. Nat Neurosci 2016;16(4):533–41.
[16] Ban T. The septo-preoptico-hypothalamic system and its autonomic function. Prog Brain Res 1966;21:1–43.
[17] Hatton GI. Function-related plasticity in hypothalamus. Annu Rev Neurosci 1997;20(1):375–97.
[18] Cavadas C, Aveleira CA, Souza GF, Velloso L. The pathophysiology of defective proteostasis in the hypothalamus—from obesity to ageing. Nat Rev Endocrinol 2016;12:723–33.
[19] Groiss SJ, Ugawa Y. Cerebellar stimulation in ataxia. Cerebellum 2012;11(2):440–2.
[20] Lee SK, Wolfe SW. Peripheral nerve injury and repair. J Am Acad Orthop Surg 2000;8(4):243–52.
[21] Low PA. Autonomic nervous system function. J Clin Neurophysiol 1993;10(1):14–27.
[22] Cannon WB. Bodily changes in pain, hunger, fear, and rage: an account of recent researches into the function of emotional excitement. New York, NY: D. Appleton and Company; 1915.
[23] Gordon T, Gordon K. Nerve regeneration in the peripheral nervous system versus the central nervous system and the relevance to speech and hearing after nerve injuries. J Commun Disord 2010;43(4):274–85.
[24] Herculano-Houzel S. The remarkable, yet not extraordinary, human brain as a scaled-up primate brain and its associated cost. Proc Natl Acad Sci 2012;109(Suppl 1):10661–8.
[25] SRy C. Studies on vertebrate neurogenesis. Scotland: Thomas; 1960.
[26] Martínez-Cerdeño V. Dendrite and spine modifications in autism and related neurodevelopmental disorders in patients and animal models. Dev Neurobiol 2016;77(4):393–404.
[27] White JA, Banerjee R, Gunawardena S. Axonal transport and neurodegeneration: how marine drugs can be used for the development of therapeutics. Mar Drugs 2016;14(5):102.
[28] Baumann N, Pham-Dinh D. Biology of oligodendrocyte and myelin in the mammalian central nervous system. Physiol Rev 2001;81(2):871–927.
[29] Arancibia-Carcamo IL, Attwell D. The node of Ranvier in CNS pathology. Acta Neuropathol 2014;128(2):161–75.
[30] Du Y, Dreyfus CF. Oligodendrocytes as providers of growth factors. J Neurosci Res 2002;68(6):647–54.
[31] Ornelas IM, McLane LE, Saliu A, Evangelou AV, Khandker L, Wood TL. Heterogeneity in oligodendroglia: is it relevant to mouse models and human disease? J Neurosci Res 2016;94(12):1421–33.
[32] Goldman SA, Kuypers NJ. How to make an oligodendrocyte. Development 2015;142(23):3983–95.
[33] Raff M, Abney E, Cohen J, Lindsay R, Noble M. Two types of astrocytes in cultures of developing rat white matter: differences in morphology, surface gangliosides, and growth characteristics. J Neurosci 1983;3(6):1289–300.

[34] Ffrench-Constant C, Raff MC. Proliferating bipotential glial progenitor cells in adult rat optic nerve. Nature 1985;319(6053):499–502.
[35] Nave K-A, Werner HB. Myelination of the nervous system: mechanisms and functions. Annu Rev Cell Dev Biol 2014;30:503–33.
[36] Stys PK, Zamponi GW, van Minnen J, Geurts JJ. Will the real multiple sclerosis please stand up? Nat Rev Neurosci 2012;13(7):507–14.
[37] Asterias Biotherapeutics. Asterias Biotherapeutics receives safety clearance to begin administering the highest dose of AST-OPC1 in the SCiStar phase 1/2a clinical trial in cervical spinal cord injury patients, 2016.
[38] Sofroniew MV, Vinters HV. Astrocytes: biology and pathology. Acta Neuropathol 2010;119(1):7–35.
[39] Abbott NJ, Rönnbäck L, Hansson E. Astrocyte–endothelial interactions at the blood–brain barrier. Nat Rev Neurosci 2006;7(1):41–53.
[40] Nulty J, Alsaffar M, Barry D. Radial glial cells organize the central nervous system via microtubule dependant processes. Brain Res 1625;2015:171–9.
[41] Xia X, Ahmad I. Unlocking the neurogenic potential of mammalian Müller glia. J Stem Cells 2016;9(2):169–75.
[42] Pekny M, Pekna M. Astrocyte reactivity and reactive astrogliosis: costs and benefits. Physiol Rev 2014;94(4):1077–98.
[43] Neumann M, Müller V, Görner K, Kretzschmar HA, Haass C, Kahle PJ. Pathological properties of the Parkinson's disease-associated protein DJ-1 in α-synucleinopathies and tauopathies: relevance for multiple system atrophy and Pick's disease. Acta Neuropathol 2004;107(6):489–96.
[44] Chen C, Chan A, Wen H, Chung S-H, Deng W, Jiang P. Stem and progenitor cell-derived astroglia therapies for neurological diseases. Trends Mol Med 2015;21(11):715–29.
[45] Mohn TC, Koob AO. Adult astrogenesis and the etiology of cortical neurodegeneration. J Neuropathol Exp Neurol 2015;9(Suppl 2):25.
[46] Kim HA, Mindos T, Parkinson DB. Plastic fantastic: Schwann cells and repair of the peripheral nervous system. Stem Cells Transl Med 2013;2(8):553–7.
[47] Newbern JM. Chapter seven-molecular control of the neural crest and peripheral nervous system development. Curr Top Dev Biol 2015;111:201–31.
[48] Arthur-Farraj PJ, Latouche M, Wilton DK, Quintes S, Chabrol E, Banerjee A, et al. C-Jun reprograms Schwann cells of injured nerves to generate a repair cell essential for regeneration. Neuron 2012;75(4):633–47.
[49] Terenghi G. Peripheral nerve regeneration and neurotrophic factors. J Anat 1999;194(1):1–14.
[50] Reilly MM, Murphy SM, Laurá M. Charcot-Marie-tooth disease. J Peripher Nerv Syst 2011;16(1):1–14.
[51] Yuki N, Hartung H-P. Guillain–Barré syndrome. N Engl J Med 2012;366(24):2294–304.
[52] Klase ZA, Khakhina S, Schneider ADB, Callahan MV, Glasspool-Malone J, Malone R. Zika fetal neuropathogenesis: etiology of a viral syndrome. PLoS Negl Trop Dis 2016;10(8):e0004877.
[53] Kanno H, Pearse DD, Ozawa H, Itoi E, Bunge MB. Schwann cell transplantation for spinal cord injury repair: its significant therapeutic potential and prospectus. Rev Neurosci 2015;26(2):121–8.
[54] Ramón-Cueto A, Avila J. Olfactory ensheathing glia: properties and function. Brain Res Bull 1998;46(3):175–87.

[55] Chou R-H, Lu C-Y, Fan J-R, Yu Y-L, Shyu W-C. The potential therapeutic applications of olfactory ensheathing cells in regenerative medicine. Cell Transplant 2014;23(4–5):567–71.
[56] Prinz M, Tay TL, Wolf Y, Jung S. Microglia: unique and common features with other tissue macrophages. Acta Neuropathol 2014;128(3):319–31.
[57] Silver J, Schwab ME, Popovich PG. Central nervous system regenerative failure: role of oligodendrocytes, astrocytes, and microglia. Cold Spring Harb Perspect Biol 2015;7(3):a020602.
[58] Ransohoff RM. How neuroinflammation contributes to neurodegeneration. Science 2016;353(6301):777–83.
[59] Davis AA, Temple S. A self-renewing multipotential stem cell in embryonic rat cerebral cortex. Nature 1994;372(6503):263–6.

CHAPTER

Stem cells and their applications in repairing the damaged nervous system

3

1 INTRODUCTION

Searching Google for the term "stem cells" returns ~77,000,000 results as of Spring 2017, indicating the high level of interest in these special cells from across the population. The potential of stem cells to treat a wide variety of diseases has captured the imagination of the general public, in addition to the minds of the scientists and engineers who study and harness their unique properties for such applications. Stem cells are often portrayed in popular culture as magic cures for ailments ranging from aging as seen in Season 3 of American Horror Story where a witch confronts a scientist at the University of Wisconsin about obtaining more stem cells, to mysterious diseases as seen in the popular TV show Orphan Black where the protagonists seek to use embryonic stem cells to cure a disease that afflicts cloned human. While these depictions of potential applications for stem cells may seem far-fetched, there is a bit of truth in these portrayals as stem cells could potentially cure several diseases, including those of the nervous system. This section defines what stem cells are and details the different types of stem cells relevant to the nervous system. The rest of this chapter then examines how these different types of stem cells could help restore neural function lost to disease or injury.

Stem cells have garnered significant attention from both the scientific community and the public due to their two unique and defining properties [1]. First, a stem cell must be able to produce other types of cells. The process by which stem cells transform into these matured cell types is termed differentiation, as the cell differentiates itself from its previous state of stemness. While the human body does an excellent job of directing stem cells to form different types of tissues during development, being able to control the process of differentiation outside of the body remains quite challenging. The second defining property of a stem cell is the ability to self-renew or replicate, meaning that it can produce more stem cells. This property makes these cells incredibly valuable for cell therapy applications as most mature cells cannot replicate, including many of the cells found in the nervous system. Often times, large quantities of cells are needed for therapeutic applications, such as the recent clinical trial by Asterias Biotherapeutics where 10 million cells were transplanted into the injured spinal cord of each patient [2]. Thus, being able to expand the number of stem cells enables scientists to grow the desired number of cells before differentiating them into the appropriate cell types necessary for therapeutic applications. These two unique properties contribute to the excitement around using stem cells for clinical

Engineering Neural Tissue from Stem Cells. http://dx.doi.org/10.1016/B978-0-12-811385-1.00003-0

applications in cell therapy. Long before Dr. Ernst McCulloch and Dr. James Till successfully identified stem cells in the 1960s in Toronto, scientists speculated about the concept of stem cells that could produce other cell types while proliferating [3]. Fig. 1 shows an example of such speculation by two different scientists in the late 1800s and early 1900s about what form a stem cell would take. Both images show how such a stem cell could divide and differentiate into other cells to create tissues. In fact, scientists today currently perform lineage tracing experiments to better understand stem cell behavior and how they differentiate them into mature cell types [4].

Many different types of stem cells exist. Each individual type of stem cell has its own unique properties. Broadly, stem cells can be classified as (1) pluripotent, (2) fetal, or (3) adult, depending on how they were derived and their unique properties. The property of pluripotency indicates that a stem cell can produce any of the cell types found in an organism. Two types of pluripotent stem cells exist—embryonic stem cells and induced pluripotent stem cells [5]. Accordingly, human pluripotent stem cells can produce cells from any tissue present in the human body. A term that is closely related to pluripotent is totipotent. Totipotent cells are pluripotent and they possess an important additional property. They can also give rise to the cells found in the placenta, which makes them distinct from pluripotent stem cells. Pluripotent stem cells can theoretically generate replacements for any tissue found in the human body, resulting in widespread excitement and research into their use. In addition to pluripotent and totipotent stem cells, other types of stem cells reside in the tissues of the human body. These populations provide an endogenous mechanism for repairing damaged tissue after disease or injury. These tissue-specific stem cells are further classified based on the age of the tissue where they reside. For example, fetal stem cells are isolated from tissue obtained before an organism is born [6,7]. As these cells actively contribute to the development of a living organism, they tend to have a higher regenerative capacity and higher levels of proliferation compared to stem cells isolated from the same tissue in an adult organism [8]. Tissue-specific stem cells tend to give rise to a limited range of cell phenotypes, and these cell types are usually restricted to the cells present in the tissue where the stem cells are located. Fig. 2 shows examples of how neural stem cells and pluripotent stem cells can be used to treat disorders of the nervous system while Fig. 3 details all the types of stem cells that can potentially give rise to neurons.

As mentioned earlier, there is enormous interest in tapping into the potential of stem cells to treat diseases and disorders. In addition, many scientists use stem cells as a tool for understanding the basic biological mechanisms behind the development of the body as well as to study diseases of the nervous system. Interestingly, the history of the stem cells and the major discoveries associated with them has driven both governments and other organizations to invest heavily in stem cell research. For example, the Stem Cell Network organization connects researchers from across Canada by supporting their efforts to develop stem cell therapies and hosting annual meetings for the exchange of ideas. This organization pays tribute to Canadian stem cell history by naming their annual meeting after the Canadian researchers, Drs. Till and McCulloch, who confirmed the existence of stem cells in the 1960s [9–11]. Their

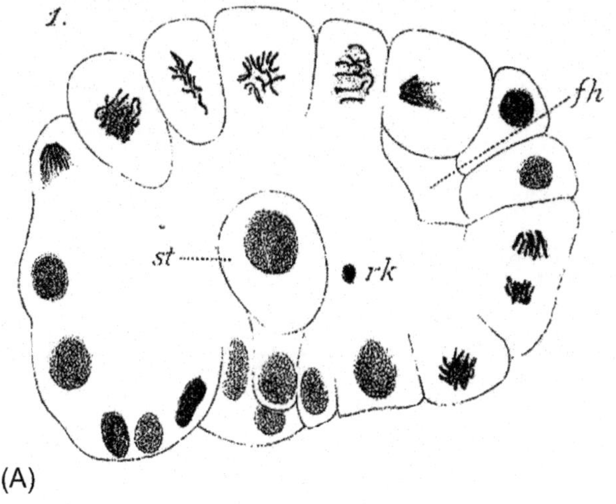

(A)

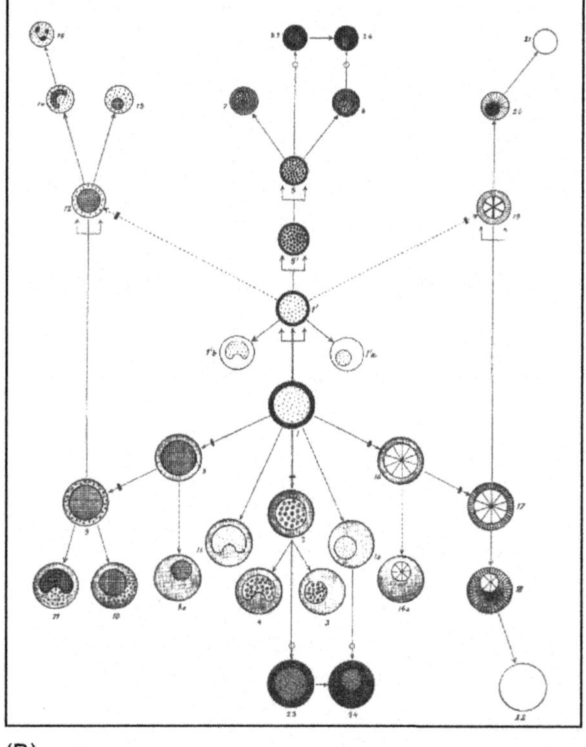

(B)

FIG. 1

Early uses of the term stem cell. (A) Valentin Häcker's 1892 diagram of the stem cell in the Cyclops embryo. According to Häcker, the stem cell (st) has just migrated to the inside of the embryo and is about to undergo asymmetric cell division. One daughter cell will give rise to the germline; the other will give rise to mesodermal tissue. (B) Artur Pappenheim's view of hematopoiesis from 1905. The cell in the center is the hypothesized common progenitor of the entire blood system. Pappenheim called this cell, among other terms, the stem cell.

Reprinted from Ramalho-Santos M, Willenbring H. On the origin of the term "stem cell." Cell Stem Cell 2007;1(1):35-8, with permission from Elsevier.

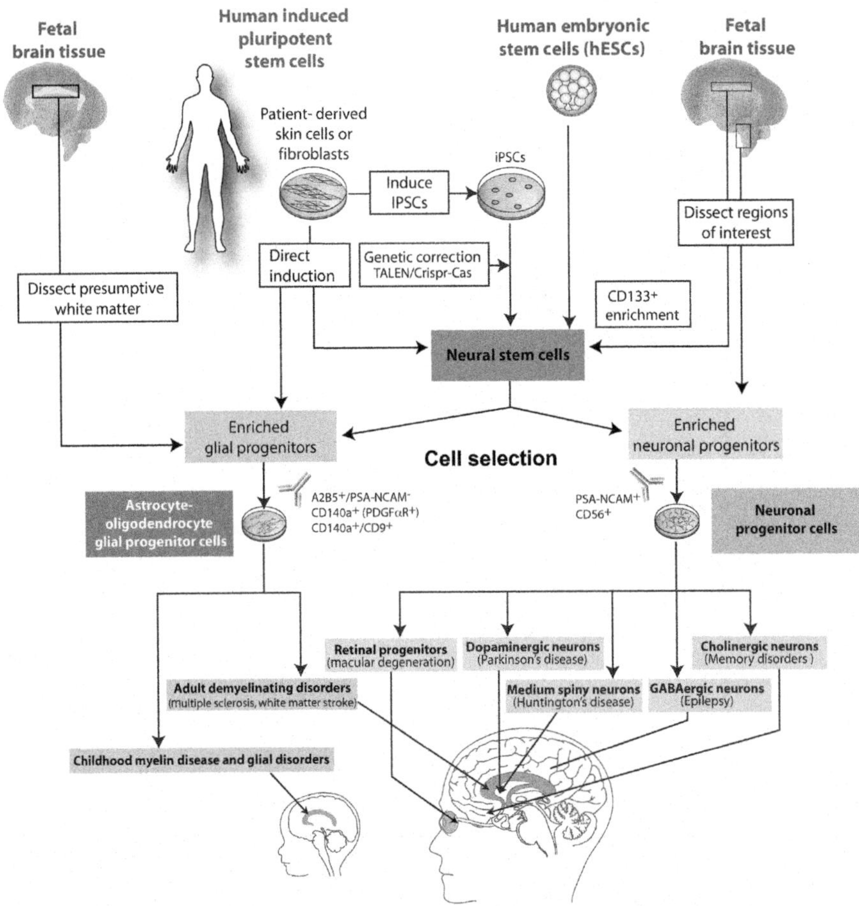

FIG. 2

Neural and glial cell therapeutics and their disease targets. This schematic illustrates the principal sources of transplantable human neural stem cells and phenotypically restricted neuronal and glial progenitor cells, and it highlights the most feasible current opportunities for their use in treating disorders of the brain.

Reprinted with permission from Goldman SA. Stem and progenitor cell-based therapy of the central nervous system: hopes, hype, and wishful thinking. Cell Stem Cell 2016;18(2):174-88.

contribution to stem cell history is discussed in the next section. In the United States, several states have established their own initiatives to promote stem cell research, with notable examples including the California Institute for Regenerative Medicine and the New York Stem Cell Foundation. The University of Wisconsin-Madison also has a rich history of pioneering stem cell research as Dr. Jamie Thompson was the first person to derive human embryonic stem cells [12]. More information on these organizations can be found at the end of this chapter. Likewise, Japan has recently

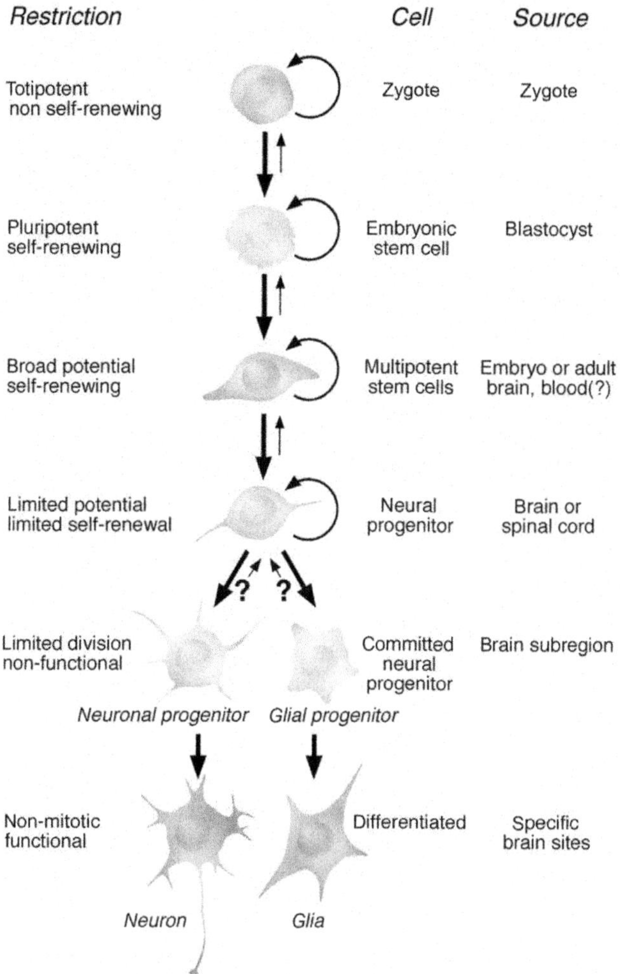

FIG. 3

An illustration proposing the classes of mammalian stem cells that can give rise to neurons, presented as a hierarchy beginning with the most primitive and multipotent stem cell and progressing to the most restricted. The restrictions of fate at each step and examples of sites in the body where they can be obtained are also presented. As our understanding of the true potential and nature of stem cells is still unfolding, modifications will clearly be added. The *small arrows* pointing up suggest the potential for dedifferentiation of the more restricted cell below.

Reprinted with permission from Gage FH. Mammalian neural stem cells. Science 2000;287(5457):1433–8.

made significant investments in developing therapies using induced pluripotent stem cells for regenerative medicine applications, inspired by the work of Nobel Prize winner—Dr. Shinya Yamanaka who invented these cells [13,14].

Interestingly, the history of stem cell biology is intertwined closely with the ability to culture cells outside of the body for sustained periods of time. As our ability to culture cells in vitro increased, it became possible to identify and study stem cells in culture dishes as opposed to using animal models where the initial stem cell studies were performed. Stem cells have intriguing similarities to the cancer stem cells, including their ability to proliferate. Being able to control proliferation is a key issue when developing stem cell therapies and understanding the mechanisms behind such control can also give insight into how cancer cells form and behave [15]. In addition to the scientific aspects of working with stem cells, their promise for cures has led patients to seek out experimental therapies using such cells to treat diseases. It is important to ensure that patients understand whether their stem cell treatments are being performed with proper approval by government regulatory bodies or if their procedure is being done without proper approval. The end of this chapter contains a link to the official list of clinical trials using stem cells currently being run in the United States. The International Society for Stem Cell Research also has extensive resources on their website for lay people who are interested in obtaining stem cell treatments, educating them on how to determine whether a stem cell clinic is operating within the proper regulatory framework as opposed to being unlicensed or a scam treatment. Such unregulated treatments often use stem cells obtained from animals and the lack of cellular characterization combined with an inability to control cell behavior post-transplantation can result in tumor formation and other negative side effects.

This chapter will now discuss the discovery and characterization of mesenchymal stem cells, which were the first type of stem cells to be identified, and how these cells can potentially be used to treat nervous system disorders. It then will move on to both fetal and adult neural stem cells followed by both types of pluripotent stem cells. Finally, it will end by introducing the concept of direct reprogramming of cells into neural tissue as an emerging area in regenerative medicine. This topic will also be revisited in Chapter 8 as an intriguing strategy for the next generation of tissue engineering.

2 DISCOVERY OF BONE MARROW STEM CELLS, THEIR ROLE IN NERVOUS SYSTEM, AND THEIR POTENTIAL APPLICATIONS IN TREATING THE DAMAGED NERVOUS SYSTEM

This section will cover the discovery of stem cells in 1960s as bone marrow stem cells were the first type of stem cell to be identified. As their name implies, they belong in the category of tissue-specific stem cells. These stem cells are now known to contain two populations of stem cells: mesenchymal stem cells and hematopoietic stem cells. Each of these stem cell populations has unique properties that will be detailed below. These stem cells are used for a wide variety of applications, including treating certain diseases of the nervous system. Their discovery, their properties, and their applications will be discussed in the following sections.

2.1 DISCOVERY OF BONE MARROW STEM CELLS AND THEIR ROLE IN THE BODY

As mentioned in the Introduction, the concept of a stem cell had been hypothesized long before their existence was confirmed in 1960. This idea of a stem cell was inspired by two different biological structures. The first structure, the embryo, is shown in Fig. 1A. The embryo contains all the instructions and cells necessary to produce a mature organism, suggesting that a small cell population gives rise to a much larger and diverse cell population. The second structure was the circulatory system. Scientists had observed the ability of the body to regenerate blood post injury, causing them to speculate on how it was replenishing itself. This observation suggested that there was a subset of cells present in the blood that could produce the rest of the cells as detailed in Fig. 1B.

These observations led to the now classical experiments performed in the 1960s that confirmed the presence of such stem cells. The interdisciplinary team discovered the presence of these cells, which would be later termed mesenchymal stem cells, while they were studying how treatment with radiation affected the bone marrow. This team consisted of Dr. Ernest McCulloch, who was a cell biologist, and Dr. James Till, a biophysicist. They were both researching how radiation could be used to treat cancer at the Ontario Cancer Institute. They observed a correlation between how sensitive bone marrow cells from healthy mice were to radiation and the amount of cells that survived transplantation [9]. This cell measurement technique could quantify the number of cells present in extracted bone marrow. They then showed a direct correlation between the number of bone marrow cells injected into an irradiated mouse with no bone marrow, and the number of nodules consisting of proliferating cells observed in the spleen [10]. Finally in 1963, they published their unifying hypothesis in Nature—indicating that these cells were, in fact, stem cells by confirming that each of the nodules arose from a single clonal cell [11].

Extensive follow-up work confirmed that bone marrow contained two distinct populations of stem cells: (1) hematopoietic stem cells and (2) mesenchymal stem cells. Both stem cell populations play important roles in the immune system, which allows our bodies to defend against foreign organisms that cause infections. Hematopoietic stem cells can recapitulate all the types of cells found in blood, including the variety of white and red blood cells required for proper function of the immune system [16]. On the other hand, mesenchymal stem cells can differentiate into other tissues such as bone, fat, and muscle tissues and they also secrete numerous growth factors that can modulate the immune system [17]. While hematopoietic stem cells can be used to treat leukemia by transplanting bone marrow [18], they are not used for treatment of the diseased or damaged nervous system. However, mesenchymal stem cells may be able to alleviate some nervous system disorders due to their unique properties that are detailed in the next section [19].

2.2 APPLICATION OF MESENCHYMAL STEM CELL THERAPIES FOR APPLICATIONS IN THE NERVOUS SYSTEM

As mentioned in the previous section, mesenchymal stem cells are multipotent as they can differentiate in adipocytes (fat cells), chondrocytes (cells found in cartilage), and osteoblasts (bone cells) [20]. They are derived from stromal (connective) tissue and express the following cell surface markers CD105, CD90, and CD73 [21]. While there have been reports that mesenchymal stem cells can differentiate into neuron-like cells [22], it is unclear if these stem cells can generate truly functional neurons in terms of their electrophysiological properties and neurotransmitter release [23]. While their ability to generate functional neurons remains unclear, the ability of mesenchymal stem cells to secrete a variety of growth factors, known as the "secretome," makes them attractive for treating neural disorders [24]. In addition, the ability to isolate mesenchymal stem cells from bone marrow makes it possible to transplant these cells autologously, which means no immunosuppression would be necessary during and after the transplantation process. Also, these cells do not pose the risk of tumor formation associated with pluripotent stem cells [25]. In terms of clinical translation, mesenchymal stem cells have been evaluated in a number of clinical trials for a wide variety of applications, ranging from treating bone disorders to diabetes [26]. In terms of neurological disorders, clinical trials have evaluated mesenchymal stem cells for their ability to treat multiple sclerosis and spinal cord injury, but these trials did not advance further due to lack of results [27]. The most promising applications of mesenchymal stem cells appear to be in the treatment of Huntington's disease and amyotrophic lateral sclerosis. The secretion of growth factors by mesenchymal stem cells alleviates the symptoms of Huntington's in animal models by creating a neuroprotective environment [28,29]. There are plans to translate such cell therapies into clinical trials. Mesenchymal stem cells exhibit similar neuroprotective effects when used to treat amyotrophic lateral sclerosis in preclinical models as well [30]. Other applications for using mesenchymal stem cell-based therapies include stroke, spinal cord injury, brain injury, and Parkinson's disease [31,32]. Overall, the ease of autologous transplantation makes it easier to evaluate the potential therapeutic effect of mesenchymal stem cells in the clinical setting. However, the effectiveness of these cells will depend on the application and the strength of their neuroprotective abilities.

3 DISCOVERY OF NEURAL STEM CELLS AND THEIR APPLICATIONS FOR NERVOUS SYSTEM REPAIR

Neural stem cells are tissue-specific stem cells found in both the fetal and adult nervous systems [33]. Their multipotent nature means they can become functional neurons as well as glial cells, including oligodendrocytes and astrocytes. They can also proliferate in culture and form structures called neurospheres, aggregates of stem cells that are beginning to differentiate into more mature neural cells [34]. In addition, the formation of neurospheres can confirm that cells isolated from the nervous system are true neural stem cells that exhibit both the ability to proliferate and differentiate. Traditionally, neural stem cells were isolated from embryonic or fetal tissue with the isolated cells often referred to as radial glial cells [35].

In the late 1980s and early 1990s, it was confirmed that adult neural stem cells were found in specific regions of the brain detailed in Chapter 2 and recapped in a recent review [36]. Dr. Sally Temple demonstrated that such neural stem cells resided in the adult mouse brain in 1989 [37], which was soon followed by the confirmation that they existed in the adult human brain by Reynolds and Weiss in 1992 [34]. While the concept of neurogenesis in the adult brain had been previously described by Altman and Das in 1965 [38,39], it took until the early 1990s to gain widespread acceptance by the scientific community. My group, in collaboration with the Swayne lab also at the University of Victoria, wrote an open-access article for the general public about the discovery of neural stem cells in the adult brain published in the journal Frontiers for Young Minds, which is listed in the resources at the end of the chapter. This paper provides a lay summary on these cells and their potential for both children and the public. In addition to the neural stem cells located in the central nervous system, the developing peripheral nervous system contains progenitors known as neural crest stem cells [40]. The neural crest forms during development in the region between the neural tube and the skin. It is considered to be part of the neuroepithelium, a thin layer of tissue found in the nervous system. These cells can produce a variety of tissues, including muscle and nerves. If we were to consider the range of stem cells as a spectrum, neural crest cells would fall somewhere in between mesenchymal stem cells and neural stem cells.

Neural and neural crest stem cells provide an attractive option for cell transplantation as they only give rise to neural cells, making it easier to generate neural tissue compared to pluripotent stem cells. This limited differentiation potential also reduces the chance of tumor formation. However, there is some evidence that neural stem cells may play a role in the formation of glioblastomas (tumors in the brain that arise from astrocytes), although it is unclear how to address this connection therapeutically [41]. In terms of disadvantages, these cells do not proliferate as extensively as pluripotent stem cells, making it challenging to obtain the large numbers of cells required for therapeutic applications. To address this issue, scientists have generated immortalized neural stem cells, where they use viral mediated methods to enable these cells to continuously proliferate [42].

Exogenous transplantation of neural stem cells has been evaluated for a range of nervous system disorders as detailed in the following recent reviews [43,44]. Most these studies have been performed at the preclinical stage. In terms of treating patients, the company Neuralstem has ongoing clinical trials evaluating the effect of their proprietary neural stem cell line—NSI-566 [45]—on the alleviating effects of amyotrophic lateral sclerosis, spinal cord injury, and stroke (http://www.neuralstem.com/patient-info-treatments-in-development) [46,47]. An alternative approach to cell therapy delivers cues to endogenous neural stem cells resident in the brain to stimulate them into repairing damage to the nervous system [48]. In both types of cell therapy, the goal is to use these neural stem cells to replace the functions lost after disease or injury. Overall, neural stem cells hold significant promise for treating neural diseases and disorders.

4 DERIVATION OF HUMAN EMBRYONIC STEM CELLS AND THEIR APPLICATIONS FOR NERVOUS SYSTEM REPAIR

The process of human development starts when a sperm cell fuses with an egg during fertilization, producing a single cell called a zygote. While sperm and eggs are haploid, meaning they each contain one set of chromosomes, the resulting zygote is diploid, containing a duplicate set of chromosomes like the rest of the somatic cells found in the human body. The zygote then begins to grow and develop, first into an embryo for the first 12 weeks following fertilization before becoming a fetus. These processes all occur while being gestated inside of the mother. As the zygote eventually produces all the cells present in the body, scientists hypothesized that cells present at the initial stages of development would consist of a special type of stem cell that possesses the property of pluripotency. The presence of these pluripotent stem cells in mammalian embryos was confirmed in 1981 when mouse embryonic stem cells were successfully isolated from the inner cell mass of a blastocyst, a structure that forms soon after the zygote is fertilized [49,50]. Due to the controversy surrounding the use of human embryos for scientific research, it took until 1998 for Jamie Thomson and colleagues at the University of Wisconsin-Madison to successfully derive human embryonic stem cells [12]. Embryonic stem cells are often cultured as aggregates known as colonies as seen in Fig. 4, which shows a colony of human embryonic stem cells. An excellent review by Davor Solter discusses the history of how these cells were derived and their relationship to teratocarcinomas, cancers

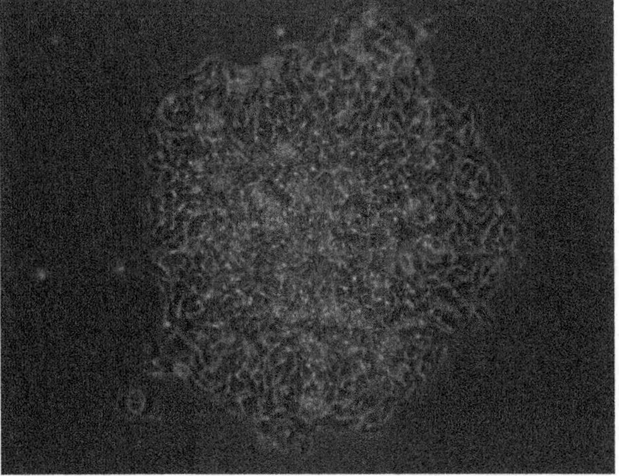

FIG. 4

Human embryonic stem cell colony cultured in all chemically defined conditions. The cells shown in the figure are the H9 cell line.

Image taken by Willerth lab.

associated with development [51]. In fact, these embryonic stem cells lines were often derived using media conditioned by cancer stem cell lines, illustrating the similarities between these two types of cells in terms of culture conditions and factors that influence their behavior.

Pluripotency provides one of the major advantages of using embryonic stem cells compared to other types of stem cells. Scientists were able to generate a lot of public attention round these cells as they can become any type of cell found in the body, and thus they have the potential to cure any disease or disorder of the body involving dysfunctional cells [52]. In addition, their ability to self-replicate provides a natural mechanism for generating sufficient quantities of cells for therapeutic applications. The generation of human embryonic stem cells also enabled the development of disease models for studying how cells behave with different types of genetic mutations associated with neural diseases [53]. Such studies were not possible before the development of embryonic stem cells as adult neural cells taken from patients suffering from such diseases had low rates of proliferation. Such adult cell lines also do not allow for studying events that occur during development. However, despite these attractive features and their potential to revolutionize cell therapy, the use of human embryonic stem cells remains controversial as the derivation of these cell lines required the destruction of embryos [54]. In particular, the sourcing of embryos was controversial. George Bush placed restrictions on preventing the use of government money on deriving new human embryonic stem cells during his presidency, which was later reversed by President Barak Obama [55]. However, attention was shifted away from human embryonic stem cells and their sourcing with the development of induced pluripotent stem cells.

Besides the controversy over their derivation, human embryonic stem cells also possess other disadvantages when it comes to applications in cell therapy. Being able to differentiate consistently human embryonic stem cells into the desired cell types remain a challenge. Their pluripotent state also means it takes them longer to mature when compared to tissue-specific stem cells. Also, their ability to proliferate continuously can lead to the formation of tumors postimplantation [56], a major concern when transplanting these cells in patients. Many of the original human embryonic stem cell lines were derived in the presence of animal products such as fetal bovine serum and Matrigel. Antigens derived from these animal products could then be expressed by the human embryonic stem cell lines [57]. Such proteins would trigger an immune response if these cells were transplanted into humans. Matching human embryonic stem cell lines to patients requires banking of cell lines that correspond to the different human leucocyte antigens (HLAs) to reduce the risk of rejection [58], which is why several countries have established national stem cell banks to store diverse cell lines.

Despite the aforementioned challenges, the first clinical trial to use cells derived from human embryonic stem cells started enrolling patients in 2010 [59]. This study was initially conducted by the Geron Corporation and their cell therapy was transplanted into patients suffering from chronic spinal cord injury. This trial was based on several preclinical studies showing that oligodendrocytes derived from human embryonic stem cells could promote functional recovery in chronic models of spinal cord injury [60–65]. The initial trial was halted 2011 due to a lack of funding despite promising results showing cell tolerance. Asterias Biotherapeutics then bought the assets associated with the trial in 2013 [66]. They continued with clinical trials and recently reported promising findings, indicating that their cell therapy promoted recovery in their patients [2]. Since this ground-breaking clinical trial, human embryonic stem cell-derived retinal pigment epithelium cells have also been evaluated in patients suffering from macular degeneration [67,68]. These transplanted cells improved vision, although the mechanism of recovery is unclear [69]. These trials are further expanding at a global level, including North America, Europe, and Asia. These clinical results indicate the potential of human embryonic stem cells for treating diseases and disorders of the nervous system. While such therapies may soon be evaluated in the clinic for treatment of Parkinson's disease [70], cell therapy using human embryonic stem cells for other applications, like Alzheimer's disease, will require more preclinical testing before translation [71].

5 DERIVATION OF INDUCED PLURIPOTENT STEM CELLS AND THEIR APPLICATIONS FOR NERVOUS SYSTEM REPAIR

The concept of cellular reprogramming, where one type of cell becomes converted into an entirely different phenotype, has been around for a long time. In 1962, Dr. John Gurdon showed that taking the nucleus isolated from a somatic cell taken from a tadpole and transplanting it into an embryo resulted in the DNA being programmed back to an embryonic state [72]. The work demonstrated the first successful example of cellular reprogramming and the process used to achieve reprogramming is called somatic cell nuclear transfer. Somatic cell nuclear transfer was later used to generate clones for a variety of animals, including the famous example of Dolly the sheep [73].

In more recent work related to stem cells, Dr. Shinya Yamanaka and his colleague Dr. Kazutoshi Takahashi screened numerous transcription factors to determine if a combination of these factors could turn somatic cells into cells that behaved like embryonic stem cells. Dr. Yamanaka first presented his ground-breaking work at the 2006 International Society for Stem Cell Research meeting held in Toronto. Their initial work used viral expression of four transcription factors, Oct4, Sox2, cMyc, and Klf4, to convert mouse fibroblasts into pluripotent stem cells that could proliferate and generate cells from all lineages [13]. Follow-up work by several research groups improved the efficiency of the reprogramming process and generated induced pluripotent stem cells that could give rise to chimeric mice, indicating that these cells had the potential to recapitulate an entire organism and providing definitive proof of pluripotency [74–76]. These cells were termed induced pluripotent stem cells. Takahashi and Yamanaka quickly followed up this work by showing that the same process could generate

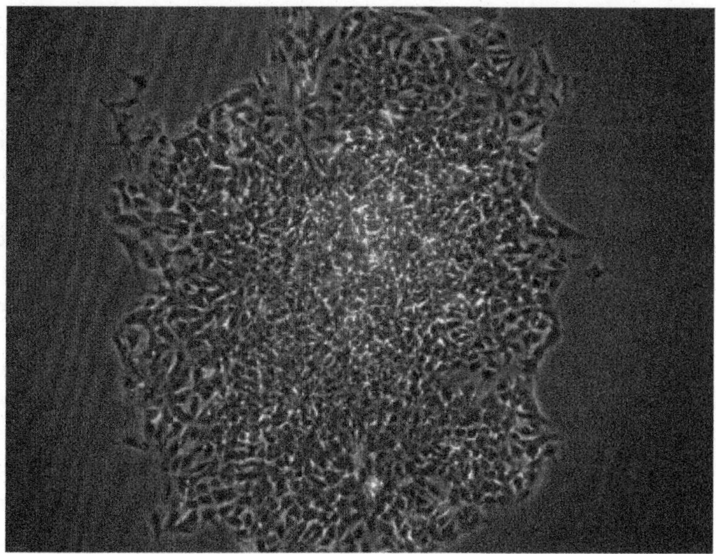

FIG. 5

Human-induced pluripotent stem cell colony (1-DL-01 cell line from WiCell) cultured in defined conditions.

Image taken by the Willerth lab.

pluripotent stem cells from human fibroblasts, which was confirmed simultaneously by the Thomson group [14,77]. Fig. 5 shows a colony of undifferentiated human-induced pluripotent stem cells, which look remarkably similar to colonies of human embryonic stem cells. Numerous follow-up studies confirmed the robustness of the reprogramming process and characterized how similar induced pluripotent stem cells were to human embryonic stem cells. The ability to generate pluripotent stem cells from somatic cells was truly a game changer and it was recognized as such in 2012 when Dr. Yamanaka and Dr. Gurdon were awarded the Nobel Prize in Medicine for their work in the field of reprogramming.

The generation of human-induced pluripotent stem cells provided an attractive method for generating pluripotent stem cell lines without the controversial destruction of embryos, making them of high interest for applications in regenerative medicine. These cells could also be produced directly from somatic cells isolated from patients who needed stem cell transplants, eliminating the need for immunosuppression posttransplantation. In addition, cells taken from patients suffering from neurological diseases such as Parkinson's and Alzheimer's could be reprogrammed into pluripotent stem cells [78–80]. These cell lines are then used to study disease progression as well as potential treatments for these diseases, making them an important tool for understanding basic neuroscience and development.

Both induced pluripotent stem cells and embryonic stem cells share the risk of tumor formation. However, there is some evidence that induced pluripotent stem cell lines can be selected to have a reduced risk of tumor formation [81]. In addition, the use of viral vectors to induce the expression of the transcription factors necessary to restore pluripotency can result in unintended genomic integration, affecting cell function. A 2012 review by Puri and Nagy does a nice job of elucidating these major differences between embryonic stem cells and induced pluripotent stem cells [82]. In addition, the reprogramming efficiency of converting somatic cells to pluripotent stem cells remains low. Often cells are not completely reprogrammed, which negatively influences their differentiation potential. Ideally, new methods for reprogramming would be discovered that would increase both the efficiency and the effectiveness of this process. Examples of such methods include using non-integrating vectors [83] and using proteins to reprogram somatic cells into pluripotent stem cells [84,85].

Japan leads the world in terms of translating cell therapies derived from induced pluripotent stem cells for clinical applications. In 2014, the first clinical trial using therapies derived from human-induced pluripotent stem cells was started in Japan. This trial evaluated the use of retinal pigment epithelial cells derived from human-induced pluripotent stem cells for the treatment of macular degeneration. The induced pluripotent stem cell lines used were obtained by reprogramming the somatic cells taken from the patient who would receive the transplanted cells. However, this trial was shut down in 2015 over concerns about the quality of the cell lines being generated by reprogramming as they exhibited genetic abnormalities, despite reporting that positive effects on vision were observed in the first patient to receive the therapy [86]. This trial is anticipated to restart soon, but the resumed trial will use differentiated cells produced from allogenic cell lines as opposed to cell lines derived directly from the patient receiving the transplant. This trial demonstrates the potential of these cells for clinic applications and it also has started the task of figuring out the optimal ways to use induced pluripotent stem cell therapies for clinical applications.

6 DIRECT REPROGRAMMING OF MATURE CELLS INTO FUNCTIONAL NEURAL CELLS

The idea of reprogramming cells into different phenotypes became popular again after the discovery of induced pluripotent stem cells. It also initiated scientists to ask if directly reprogramming one mature cell type into another mature phenotype was possible. The answer was yes, and now numerous examples of this type of direct reprogramming have been published in the literature [87–93]. The reprogramming process works by changing the protein expression pattern inside of a cell to reflect the characteristics of the desired cell type. The proteins used during this process are often transcription factors, sharing similarities with the process used to generate induced pluripotent stem cells. For neural tissue engineering, studying the transcription factor networks expressed in developing nervous system can identify which factors could reprogram other mature cells into neural phenotypes.

For example, the transcription factor Ascl1 serves as a powerful regulator of developmental neurogenesis, controlling the behavior of neural stem cells [94]. Knocking out Ascl1 expression reduces neurogenesis and causes dysfunction in the nervous system [95], demonstrating its important function. Viral expression of Ascl1 directly converts adult fibroblasts into neural lineages, attracting a significant amount of attention [96–104]. Fig. 6 shows how such reprogramming works to convert mouse fibroblasts into functional neurons. In 2014, the Wernig group, one of the pioneers in the field of direct reprogramming, demonstrated that direct conversion of fibroblasts into neurons was possible only by using a lentiviral vector encoding Ascl1 [96]. Later work demonstrated that Ascl1 directly binds condensed chromatin where it can recruit other transcription factors, making it a "pioneer" transcription factor due to its potency in reprogramming [105]. The most effective methods for reprogramming human fibroblasts into neurons rely on a combination of transcription factors and the use of multiple transcription factors can specify what neuronal subtypes are generated by reprogramming. These transcription factor combinations include the BAM factors (Brn2, Ascl1, and Mytl1) [100] and the LAN factors (Lmx1a, Ascl1, and Nurr1) [106]. Other methods for reprogramming human fibroblasts into functional neurons include expressing microRNAs using viral methods [107–109] and using CRISPR/Cas9 technology to target the BAM factors [110]. Gene editing will also be discussed in more depth in Chapter 8. However, the use of viral vectors to induce protein expression remains problematic for clinical applications. The infection process alters the genome of the infected cells and produces inconsistent levels of protein expression due to the heterogeneous rates of infection [92,111,112]. As an alternative, several groups have used small molecule cocktails to reprogram fibroblasts into different types of cells found in the nervous system, neural stem cells, astrocytes (neural support cells) and neurons [113–117]. However, their mechanism of reprogramming remains unknown. Other major limitations associated with using small molecules cocktails include achieving proper dosing and the potential for off-target effects in vivo [118].

The end goal for reprogramming would be to apply this process to neural diseases and disorders in vivo to restore lost function [93]. Several groups have converted various cell types in vivo into the desired phenotypes for engineering neural tissue [104,119–123]. Niu et al. showed that the viral expression of Sox2 converted astrocytes in the brain into neural progenitors that matured into functional neurons [119]. They also later demonstrated that the overexpression of NeuroD converts reactive astrocytes present after brain injury into neurons [120]. Both human fibroblasts and astrocytes can be reprogrammed into neurons using similar methods [121]. Follow-up work confirmed that these reprogrammed cells integrated into the neural circuitry of

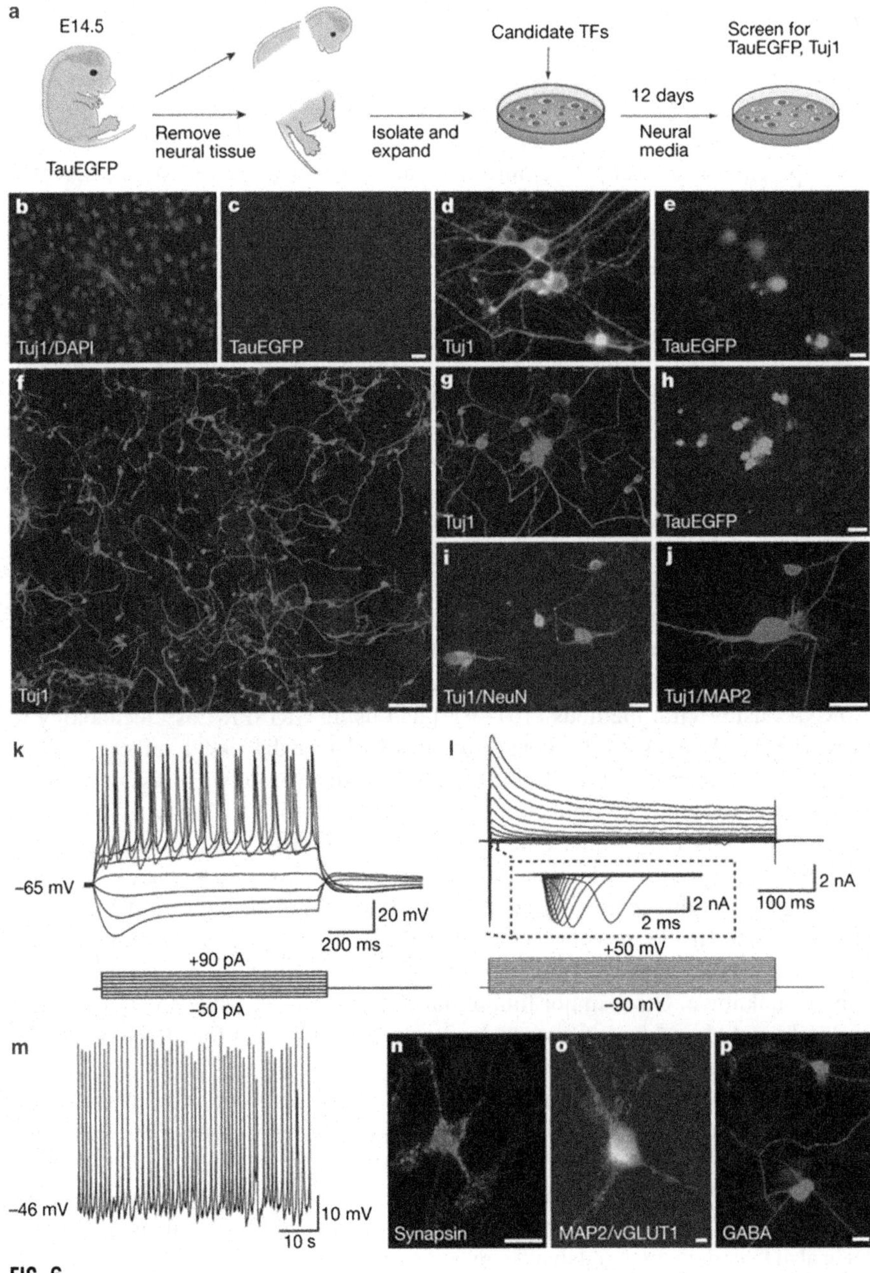

FIG. 6

See the legend on opposite page

the brain [122]. In 2015, Liu and colleagues used adeno-associated virus (AAV) to express Ascl1 in astrocytes found in the brain, converting these cells into functional neurons [123]. Successful cellular reprogramming can be achieved in the injured spinal cord as well. For example, viral overexpression of Sox2 in combination with valproic acid treatment converted the astrocytes present in the injured spinal cord into neural progenitors [124]. These neural progenitors then differentiated into mature neurons, but no assessment of functional recovery was performed.

The major advantages of these methods are that they avoid the need for exogenous cell culture, transplantation and the need for immunosuppression. Direct reprogramming can also be used to generate neural cells from patients suffering from disease, which can then be used to learn more about the disease or for drug screening applications [125]. The major challenge is overcoming the issues associated with current methods of reprogramming detailed above before translating in vivo engineering to the clinic. While in vivo examples have been published in rodent models, human fibroblasts show lower reprogramming efficiency compared to mouse fibroblasts when using virus-mediated transcription factor expression, suggesting translation might be difficult to achieve [126]. These issues must be addressed before such strategies can be translated for clinical applications.

FIG. 6

A screen for neuronal fate inducing factors and characterization of mouse embryonic fibroblast (MEF)-derived induced neuronal (iN) cells. (A) Experimental rationale. (B) Uninfected, p3 TauEGFP MEFs contained rare Tuj1-positive cells *(red in the electronic version, gray in the print version)* with flat morphology. *Blue in the electronic version, gray in the print version*: DAPI counterstain. (C) Tuj1-positive fibroblasts do not express visible TauEGFP. (D and E) MEF-iN cells express Tuj1 *(red in the electronic version, gray in the print version)* and TauEGFP *(green in the electronic version, gray in the print version)* and display complex neuronal morphologies 32 days after infection with the 19-factor (19F) pool. (F) Tuj1 expression in MEFs 13 days after infection with the 5F pool. (G–J) MEF-derived Tuj1-positive iN cells co-express the pan-neuronal markers TauEGFP (H), NeuN (*red in the electronic version, gray in the print version*, L) and MAP2 (*red in the electronic version, gray in the print version*, J). (K) Representative traces of membrane potential responding to step depolarization by current injection (lower panel). Membrane potential was current-clamped at around −65 mV. (L) Representative traces of whole-cell currents in voltage-clamp mode, cell was held at −70 mV, step depolarization from −90 to 60 mV at 10 mV intervals were delivered (lower panel). Insert showing Na^+ currents. (M) Spontaneous action potentials (AP) recorded from a 5F MEF-iN cell 8 days postinfection. No current injection was applied. (N–P) 22 days postinfection 5F MEF-iN cells expressed synapsin (*red*, N) and vesicular glutamate transporter 1 (vGLUT1) (*red in the electronic version, gray in the print version*, O) or GABA (P). Scale bars = 5 μm (O), 10 μm (E, N, P) 20 μm (C, H, I), and 200 μm (F).

Reprinted with permission from Vierbuchen T, Ostermeier A, Pang ZP, Kokubu Y, Sudhof TC, Wernig M. Direct conversion of fibroblasts to functional neurons by defined factors. Nature 2010;463(7284):1035–41.

7 CONCLUDING REMARKS ON STEM CELLS AND THEIR POTENTIAL AS CELL THERAPIES

A wide variety of stem cells can be applied to repairing the damaged nervous system as detailed in this chapter. The key events in the history of stem cell biology discussed in this chapter are summarized in two timelines shown in Fig. 7 (events occurring pre-2000s) and Fig. 8 (events occurring post-2000s). It is possible that all the stem cell lines discussed in this chapter could serve as effective cell therapies due to the complex nature of the diseases and disorders that afflict the nervous system. The ability to reprogram cells directly into functional phenotypes also

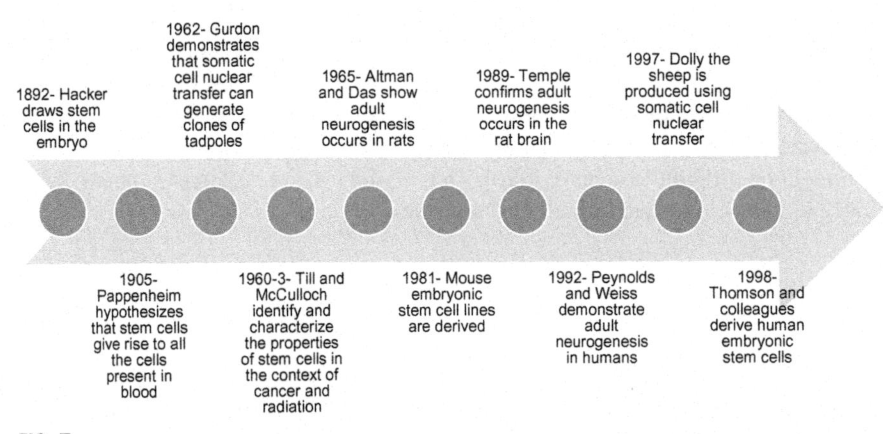

FIG. 7

Timeline of significant events in stem cell biology pre-2000s.

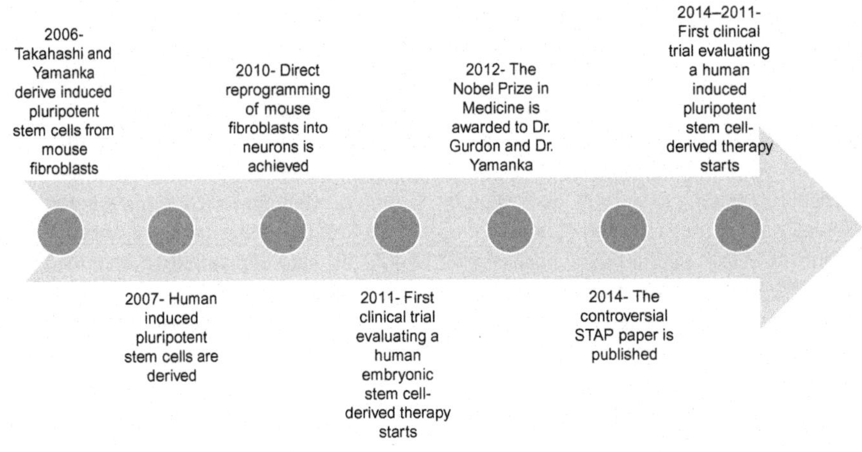

FIG. 8

Timeline of significant events in stem cell biology post-2000s.

serves as a powerful potential new method for treating such disorders. However, as with all new technology, this promise and potential must be monitored with caution to ensure that the evaluation of such therapies is properly conducted and the effects achieved are valid. A recent scandal illustrates the effect of such hype without proper scientific rigor. In 2014, Haruko Okabota and colleagues reported in a pair of Nature papers that treating somatic mouse cells with acid converted them into pluripotent stem cells [127,128]. They termed this process "stimulus-triggered acquisition of pluripotency" (STAP) and the resulting stem cells were referred to as STAP cells. However, multiple groups of scientists were unable to replicate this simple technique for generating pluripotent stem cells, raising suspicion about the data presented. Eventually, it was confirmed that the initial data presented was false [129] and the papers were retracted. The first author Okabota ended up having her Ph.D. revoked and one of her coauthors, Dr. Yoshiki Sasai, committed suicide in the aftermath of this scandal. Another coauthor Dr. Vacanti remains convinced that STAP cells do exist and has posted protocols for generating these cells on his website. The entire situation serves as an important reminder to emphasize reproducibility in science and to treat all new discoveries with a healthy degree of skepticism. However, the promising results in clinical trials detailed in this chapter suggest that we will soon see more clinical trials using stem cells to treat disorders of the nervous system.

RELEVANT RESOURCES

National Institutes of Health—Stem Cell Basics Website: https://stemcells.nih.gov/info/basics/1.htm

The Niche—Knoepfler lab blog covering current events in stem cell biology: http://www.ipscell.com/

He also provides a link to updated lists of stem cell related clinical trials by disease: http://www.ipscell.com/find-a-clinical-trial/

StemBook is an online review of stem cell biology published by the Harvard Stem Cell Institute: http://www.stembook.org/.

International Society for Stem Cell Research—hosts an annual international Stem Cell meeting: http://www.isscr.org/.

Society for Biological Engineering—a subgroup of the American Institute of Chemical Engineers—that hosts a biannual International Conference on Stem Cell Engineering: http://www.aiche.org/sbe

California Institute for Regenerative Medicine: http://cirm.ca.gov

New York Stem Cell Foundation: http://nyscf.org/

Stem Cell Network (Canada's national stem cell network): http://stemcellnetwork.ca

Link to the Super Cells Exhibit produced by the Stem Cell Network: http://www.supercells.ca/

Neuralstem Inc.—running several clinical trials evaluating neural stem cells for treating neurological disorder: http://www.neuralstem.com/

Sanchez-Arias, J., Agbay, A., Willerth, S.M., Swayne, L-A. What are "brain" stem cells and why are they important? Frontiers for Young Minds. September 2016. Frontiers for Young Minds is an open access journal for youth aged 8–15. This article was the first one on stem cells in the Neuroscience section. http://kids.frontiersin.org/article/10.3389/frym.2016.00020

REFERENCES

[1] Till JE, McCulloch EA. Hemopoietic stem cell differentiation. Biochim Biophys Acta-Rev Cancer 1980;605(4):431–59.

[2] Asterias Biotherapeutics. Asterias Biotherapeutics receives safety clearance to begin administering the highest dose of AST-OPC1 in the SCiStar phase 1/2a clinical trial in cervical spinal cord injury patients. Press release. 31st August, 2016.

[3] Ramalho-Santos M, Willenbring H. On the origin of the term "stem cell". Cell Stem Cell 2007;1(1):35–8.

[4] Chaker Z, Codega P, Doetsch F. A mosaic world: puzzles revealed by adult neural stem cell heterogeneity. Wiley Interdiscip Rev Dev Biol 2016;5(6):640–58.

[5] Willerth SM. Neural tissue engineering using embryonic and induced pluripotent stem cells. Stem Cell Res Ther 2011;2(2):17.

[6] Edwards RG. Stem cells today: A. Origin and potential of embryo stem cells. Reprod BioMed Online 2004;8(3):275–306.

[7] O'Donoghue K, Fisk NM. Fetal stem cells. Best Pract Res Clin Obstet Gynaecol 2004;18(6):853–75.

[8] Clarke DL, Johansson CB, Wilbertz J, Veress B, Nilsson E, Karlström H, et al. Generalized potential of adult neural stem cells. Science 2000;288(5471):1660–3.

[9] McCulloch EA, Till JE. The radiation sensitivity of normal mouse bone marrow cells, determined by quantitative marrow transplantation into irradiated mice. Radiat Res 1960;13:115–25.

[10] Till JE, Mc CE. A direct measurement of the radiation sensitivity of normal mouse bone marrow cells. Radiat Res 1961;14:213–22.

[11] Becker AJ, Mc CE, Till JE. Cytological demonstration of the clonal nature of spleen colonies derived from transplanted mouse marrow cells. Nature 1963;197:452–4.

[12] Thomson JA, Itskovitz-Eldor J, Shapiro SS, Waknitz MA, Swiergiel JJ, Marshall VS, et al. Embryonic stem cell lines derived from human blastocysts. Science 1998;282(5391):1145–7.

[13] Takahashi K, Yamanaka S. Induction of pluripotent stem cells from mouse embryonic and adult fibroblast cultures by defined factors. Cell 2006;126(4):663–76.

[14] Takahashi K, Tanabe K, Ohnuki M, Narita M, Ichisaka T, Tomoda K, et al. Induction of pluripotent stem cells from adult human fibroblasts by defined factors. Cell 2007;131(5):861–72.

[15] Reya T, Morrison SJ, Clarke MF, Weissman IL. Stem cells, cancer, and cancer stem cells. Nature 2001;414(6859):105–11.

[16] Eaves CJ. Hematopoietic stem cells: concepts, definitions, and the new reality. Blood 2015;125(17):2605–13.

[17] Pacini S. Deterministic and stochastic approaches in the clinical application of mesenchymal stromal cells (MSCs). Front Cell Dev Biol 2014;2:50.

[18] Pontikoglou C, Deschaseaux F, Sensebé L, Papadaki HA. Bone marrow mesenchymal stem cells: biological properties and their role in hematopoiesis and hematopoietic stem cell transplantation. Stem Cell Rev Rep 2011;7(3):569–89.
[19] Tanna T, Sachan V. Mesenchymal stem cells: potential in treatment of neurodegenerative diseases. Curr Stem Cell Res Ther 2014;9(6):513–21.
[20] Jiang Y, Jahagirdar BN, Reinhardt RL, Schwartz RE, Keene CD, Ortiz-Gonzalez XR, et al. Pluripotency of mesenchymal stem cells derived from adult marrow. Nature 2002;418(6893):41–9.
[21] Dominici M, Le Blanc K, Mueller I, Slaper-Cortenbach I, Marini F, Krause D, et al. Minimal criteria for defining multipotent mesenchymal stromal cells. The International Society for Cellular Therapy position statement. Cytotherapy 2006;8(4):315–7.
[22] Frausin S, Viventi S, Falzacappa LV, Quattromani MJ, Leanza G, Tommasini A, et al. Wharton's jelly derived mesenchymal stromal cells: biological properties, induction of neuronal phenotype and current applications in neurodegeneration research. Acta Histochem 2015;117(4):329–38.
[23] Lambert APF, Zandonai AF, Bonatto D, Machado DC, Henriques JAP. Differentiation of human adipose-derived adult stem cells into neuronal tissue: does it work? Differentiation 2009;77(3):221–8.
[24] Caplan AI, Dennis JE. Mesenchymal stem cells as trophic mediators. J Cell Biochem 2006;98(5):1076–84.
[25] Herberts CA, Kwa MS, Hermsen HP. Risk factors in the development of stem cell therapy. J Transl Med 2011;9(1):1.
[26] Squillaro T, Peluso G, Galderisi U. Clinical trials with mesenchymal stem cells: an update. Cell Transplant 2016;25(5):829–48.
[27] Wang S, Qu X, Zhao RC. Clinical applications of mesenchymal stem cells. J Hematol Oncol 2012;5(1):1.
[28] Fink KD, Deng P, Torrest A, Stewart H, Pollock K, Gruenloh W, et al. Developing stem cell therapies for juvenile and adult-onset Huntington's disease. Regen Med 2015;10(5):623–46.
[29] Kerkis I, Haddad MS, Valverde CW, Glosman S. Neural and mesenchymal stem cells in animal models of Huntington's disease: past experiences and future challenges. Stem Cell Res Ther 2015;6(1):1.
[30] Shakhbazau A, Potapnev M. Autologous mesenchymal stromal cells as a therapeutic in ALS and epilepsy patients: treatment modalities and ex vivo neural differentiation. Cytotherapy 2016;18(10):1245–55.
[31] Teixeira FG, Carvalho MM, Panchalingam KM, Rodrigues AJ, Mendes-Pinheiro B, Anjo S, et al. Impact of the secretome of human mesenchymal stem cells on brain structure and animal behavior in a rat model of Parkinson's disease. Stem Cells Transl Med 2016;6(2):634–46.
[32] Wang Y, He W, Bian H, Liu C, Li S. Small molecule induction of neural-like cells from bone marrow-mesenchymal stem cells. J Cell Biochem 2012;113(5):1527–36.
[33] Kornblum HI. Introduction to neural stem cells. Stroke 2007;38(2):810–6.
[34] Reynolds BA, Weiss S. Generation of neurons and astrocytes from isolated cells of the adult mammalian central nervous system. Science 1992;255(5052):1707.
[35] Malatesta P, Götz M. Radial glia–from boring cables to stem cell stars. Development 2013;140(3):483–6.
[36] Gage FH, Temple S. Neural stem cells: generating and regenerating the brain. Neuron 2013;80(3):588–601.

[37] Temple S. Division and differentiation of isolated CNS blast cells in microculture. Nature 1989;340(6233):471–3.
[38] Altman J, Das GD. Post-natal origin of microneurones in the rat brain. Nature 1965;207(5000):953.
[39] Altman J, Das GD. Autoradiographic and histological evidence of postnatal hippocampal neurogenesis in rats. J Comp Neurol 1965;124(3):319–35.
[40] Liu JA, Cheung M. Neural crest stem cells and their potential therapeutic applications. Dev Biol 2016;419(2):199–216.
[41] Smith AW, Mehta MP, Wernicke AG. Neural stem cells, the subventricular zone and radiotherapy: implications for treating glioblastoma. J Neuro-Oncol 2016;1–10.
[42] Villa A, Snyder EY, Vescovi A, Martínez-Serrano A. Establishment and properties of a growth factor-dependent, perpetual neural stem cell line from the human CNS. Exp Neurol 2000;161(1):67–84.
[43] Goldman SA. Stem and progenitor cell-based therapy of the central nervous system: hopes, hype, and wishful thinking. Cell Stem Cell 2016;18(2):174–88.
[44] Harris L, Zalucki O, Piper M, Heng JI-T. Insights into the biology and therapeutic applications of neural stem cells. Stem Cells Int 2016;2016:9745315.
[45] Guo X, Johe K, Molnar P, Davis H, Hickman J. Characterization of a human fetal spinal cord stem cell line, NSI-566RSC, and its induction to functional motoneurons. J Tissue Eng Regen Med 2010;4(3):181–93.
[46] Glass JD, Boulis NM, Johe K, Rutkove SB, Federici T, Polak M, et al. Lumbar intraspinal injection of neural stem cells in patients with amyotrophic lateral sclerosis: results of a phase I trial in 12 patients. Stem Cells 2012;30(6):1144–51.
[47] Tadesse T, Gearing M, Senitzer D, Saxe D, Brat DJ, Bray R, et al. Analysis of graft survival in a trial of stem cell transplant in ALS. Ann Clin Transl Neurol 2014;1(11):900–8.
[48] Herrera-Arozamena C, Martí-Marí O, Estrada M, de la Fuente Revenga M, Rodríguez-Franco MI. Recent advances in neurogenic small molecules as innovative treatments for neurodegenerative diseases. Molecules 2016;21(9):1165.
[49] Martin GR. Isolation of a pluripotent cell line from early mouse embryos cultured in medium conditioned by teratocarcinoma stem cells. Proc Natl Acad Sci U S A 1981;78(12):7634–8.
[50] Evans MJ, Kaufman MH. Establishment in culture of pluripotential cells from mouse embryos. Nature 1981;292(5819):154–6.
[51] Solter D. From teratocarcinomas to embryonic stem cells and beyond: a history of embryonic stem cell research. Nat Rev Genet 2006;7(4):319–27.
[52] Gerecht-Nir S, Itskovitz-Eldor J. Cell therapy using human embryonic stem cells. Transpl Immunol 2004;12(3):203–9.
[53] Sterneckert JL, Reinhardt P, Schöler HR. Investigating human disease using stem cell models. Nat Rev Genet 2014;15(9):625–39.
[54] Ludwig TE, Levenstein ME, Jones JM, Berggren WT, Mitchen ER, Frane JL, et al. Derivation of human embryonic stem cells in defined conditions. Nat Biotechnol 2006;24(2):185–7.
[55] Robertson JA. Embryo stem cell research: ten years of controversy. J Law Med Ethics 2010;38(2):191–203.
[56] Doi D, Morizane A, Kikuchi T, Onoe H, Hayashi T, Kawasaki T, et al. Prolonged maturation culture favors a reduction in the tumorigenicity and the dopaminergic function of human ESC-derived neural cells in a primate model of Parkinson's disease. Stem Cells 2012;30(5):935–45.

[57] Martin MJ, Muotri A, Gage F, Varki A. Human embryonic stem cells express an immunogenic nonhuman sialic acid. Nat Med 2005;11(2):228–32.
[58] Taylor CJ, Bolton EM, Pocock S, Sharples LD, Pedersen RA, Bradley JA. Banking on human embryonic stem cells: estimating the number of donor cell lines needed for HLA matching. Lancet 2005;366(9502):2019–25.
[59] Geron Corporation. World's first clinical trial of human embryonic stem cell therapy cleared. Regen Med 2009;4(2):161.
[60] Nistor GI, Totoiu MO, Haque N, Carpenter MK, Keirstead HS. Human embryonic stem cells differentiate into oligodendrocytes in high purity and myelinate after spinal cord transplantation. GLIA 2005;49(3):385–96.
[61] Keirstead HS, Nistor G, Bernal G, Totoiu M, Cloutier F, Sharp K, et al. Human embryonic stem cell-derived oligodendrocyte progenitor cell transplants remyelinate and restore locomotion after spinal cord injury. J Neurosci 2005;25(19):4694–705.
[62] Faulkner J, Keirstead HS. Human embryonic stem cell-derived oligodendrocyte progenitors for the treatment of spinal cord injury. Transpl Immunol 2005;15(2):131–42.
[63] Cloutier F, Siegenthaler MM, Nistor G, Keirstead HS. Transplantation of human embryonic stem cell-derived oligodendrocyte progenitors into rat spinal cord injuries does not cause harm. Regen Med 2006;1(4):469–79.
[64] Sharp J, Frame J, Siegenthaler M, Nistor G, Keirstead HS. Human embryonic stem cell-derived oligodendrocyte progenitor cell transplants improve recovery after cervical spinal cord injury. Stem Cells 2010;28(1):152–63.
[65] Rossi SL, Nistor G, Wyatt T, Yin HZ, Poole AJ, Weiss JH, et al. Histological and functional benefit following transplantation of motor neuron progenitors to the injured rat spinal cord. PLoS ONE 2010;5(7):e11852.
[66] Scott CT, Magnus D. Wrongful termination: lessons from the Geron clinical trial. Stem Cells Transl Med 2014;3(12):1398–401.
[67] Schwartz SD, Regillo CD, Lam BL, Eliott D, Rosenfeld PJ, Gregori NZ, et al. Human embryonic stem cell-derived retinal pigment epithelium in patients with age-related macular degeneration and Stargardt's macular dystrophy: follow-up of two open-label phase 1/2 studies. Lancet 2015;385(9967):509–16.
[68] Song WK, Park K-M, Kim H-J, Lee JH, Choi J, Chong SY, et al. Treatment of macular degeneration using embryonic stem cell-derived retinal pigment epithelium: preliminary results in Asian patients. Stem Cell Rep 2015;4(5):860–72.
[69] Hu J, He T-C, Li F. For your eyes only: harnessing human embryonic stem cell-derived retinal pigment epithelial cells to improve impaired vision. Genes Dis 2015;2(4):293–4.
[70] Lindvall O. Treatment of Parkinson's disease using cell transplantation. Philos Trans R Soc B 2015;370(1680):20140370.
[71] Tong LM, Fong H, Huang Y. Stem cell therapy for Alzheimer's disease and related disorders: current status and future perspectives. Exp Mol Med 2015;47(3):e151.
[72] Gurdon JB. The developmental capacity of nuclei taken from intestinal epithelium cells of feeding tadpoles. Development 1962;10(4):622–40.
[73] Wilmut I, Schnieke A, McWhir J, Kind A, Campbell K. Viable offspring derived from fetal and adult mammalian cells. In: Clones and clones: facts and fantasies about human cloning, vol. 21. New York: W. W. Norton & Company; 1999.
[74] Okita K, Ichisaka T, Yamanaka S. Generation of germline-competent induced pluripotent stem cells. Nature 2007;448(7151):313–7.

[75] Wernig M, Meissner A, Foreman R, Brambrink T, Ku M, Hochedlinger K, et al. In vitro reprogramming of fibroblasts into a pluripotent ES-cell-like state. Nature 2007;448(7151):318–24.
[76] Maherali N, Sridharan R, Xie W, Utikal J, Eminli S, Arnold K, et al. Directly reprogrammed fibroblasts show global epigenetic remodeling and widespread tissue contribution. Cell Stem Cell 2007;1(1):55–70.
[77] Yu J, Vodyanik MA, Smuga-Otto K, Antosiewicz-Bourget J, Frane JL, Tian S, et al. Induced pluripotent stem cell lines derived from human somatic cells. Science 2007;318(5858):1917–20.
[78] Barral S, Kurian MA. Utility of induced pluripotent stem cells for the study and treatment of genetic diseases: focus on childhood neurological disorders. Front Mol Neurosci 2016;9:78.
[79] Kriks S, Shim JW, Piao J, Ganat YM, Wakeman DR, Xie Z, et al. Dopamine neurons derived from human ES cells efficiently engraft in animal models of Parkinson's disease. Nature 2011;480(7378):547–51.
[80] Yang J, Li S, He X-B, Cheng C, Le W. Induced pluripotent stem cells in Alzheimer's disease: applications for disease modeling and cell-replacement therapy. Mol Neurodegener 2016;11(1):1.
[81] Nori S, Okada Y, Yasuda A, Tsuji O, Takahashi Y, Kobayashi Y, et al. Grafted human-induced pluripotent stem-cell-derived neurospheres promote motor functional recovery after spinal cord injury in mice. Proc Natl Acad Sci U S A 2011;108(40):16825–30.
[82] Puri MC, Nagy A. Concise review: embryonic stem cells versus induced pluripotent stem cells: the game is on. Stem Cells 2012;30(1):10–4.
[83] Woltjen K, Hämäläinen R, Kibschull M, Mileikovsky M, Nagy A. Transgene-free production of pluripotent stem cells using piggyBac transposons. Methods Mol Biol 2011;87–103.
[84] Zhou H, Wu S, Joo JY, Zhu S, Han DW, Lin T, et al. Generation of induced pluripotent stem cells using recombinant proteins. Cell Stem Cell 2009;4(5):381–4.
[85] Kim D, Kim C-H, Moon J-I, Chung Y-G, Chang M-Y, Han B-S, et al. Generation of human induced pluripotent stem cells by direct delivery of reprogramming proteins. Cell Stem Cell 2009;4(6):472.
[86] Garber K. RIKEN suspends first clinical trial involving induced pluripotent stem cells. Nat Biotechnol 2015;33(9):890–1.
[87] Heinrich C, Spagnoli FM, Berninger B. In vivo reprogramming for tissue repair. Nat Cell Biol 2015;17(3):204–11.
[88] Tanabe K, Haag D, Wernig M. Direct somatic lineage conversion. Philos Trans R Soc Lond Ser B Biol Sci 2015;370(1680):20140368.
[89] Chen G, Wernig M, Berninger B, Nakafuku M, Parmar M, Zhang C-L. In vivo reprogramming for brain and spinal cord repair. eNeuro 2015;2(5):0106–15.
[90] Ichida JK, Kiskinis E. Probing disorders of the nervous system using reprogramming approaches. EMBO J 2015;34(11):1456–77.
[91] Tsunemoto RK, Eade KT, Blanchard JW, Baldwin KK. Forward engineering neuronal diversity using direct reprogramming. EMBO J 2015;34(11):1445–55.
[92] Di Stefano B, Hochedlinger K. Cell reprogramming: brain versus brawn. Nature 2016;534(7607):332–3.
[93] Srivastava D, DeWitt N. In vivo cellular reprogramming: the next generation. Cell 2016;166(6):1386–96.
[94] Vasconcelos FF, Castro DS. Transcriptional control of vertebrate neurogenesis by the proneural factor Ascl1. Front Cell Neurosci 2014;8:412.

[95] Casarosa S, Fode C, Guillemot F. Mash1 regulates neurogenesis in the ventral telencephalon. Development 1999;126(3):525–34.

[96] Chanda S, Ang CE, Davila J, Pak C, Mall M, Lee QY, et al. Generation of induced neuronal cells by the single reprogramming factor ASCL1. Stem Cell Rep 2014;3(2):282–96.

[97] Colasante G, Lignani G, Rubio A, Medrihan L, Yekhlef L, Sessa A, et al. Rapid conversion of fibroblasts into functional forebrain GABAergic interneurons by direct genetic reprogramming. Cell Stem Cell 2015;17(6):719–34.

[98] Ueki Y, Wilken MS, Cox KE, Chipman L, Jorstad N, Sternhagen K, et al. Transgenic expression of the proneural transcription factor Ascl1 in Muller glia stimulates retinal regeneration in young mice. Proc Natl Acad Sci U S A 2015;112(44):13717–22.

[99] Gopalakrishnan S, Hor P, Ichida JK. New approaches for direct conversion of patient fibroblasts into neural cells. Brain Res 2015;1656:2–13.

[100] Vierbuchen T, Ostermeier A, Pang ZP, Kokubu Y, Sudhof TC, Wernig M. Direct conversion of fibroblasts to functional neurons by defined factors. Nature 2010;463(7284):1035–41.

[101] Marro S, Pang ZP, Yang N, Tsai MC, Qu K, Chang HY, et al. Direct lineage conversion of terminally differentiated hepatocytes to functional neurons. Cell Stem Cell 2011;9(4):374–82.

[102] Pang ZP, Yang N, Vierbuchen T, Ostermeier A, Fuentes DR, Yang TQ, et al. Induction of human neuronal cells by defined transcription factors. Nature 2011;476(7359):220–3.

[103] Niu W, Zang T, Zou Y, Fang S, Smith DK, Bachoo R, et al. In vivo reprogramming of astrocytes to neuroblasts in the adult brain. Nat Cell Biol 2013;15(10):1164–75.

[104] ZD S, Niu WZ, Liu ML, Zou YH, Zhang CL. In vivo conversion of astrocytes to neurons in the injured adult spinal cord. Nat Commun 2014;5:3338.

[105] Wapinski OL, Vierbuchen T, Qu K, Lee QY, Chanda S, Fuentes DR, et al. Hierarchical mechanisms for direct reprogramming of fibroblasts to neurons. Cell 2013;155(3):621–35.

[106] Caiazzo M, Dell'Anno MT, Dvoretskova E, Lazarevic D, Taverna S, Leo D, et al. Direct generation of functional dopaminergic neurons from mouse and human fibroblasts. Nature 2011;476(7359):224–7.

[107] Richner M, Victor MB, Liu Y, Abernathy D, Yoo AS. MicroRNA-based conversion of human fibroblasts into striatal medium spiny neurons. Nat Protoc 2015;10(10):1543–55.

[108] Zhou C, Gu H, Fan R, Wang B, Lou J. MicroRNA 302/367 cluster effectively facilitates direct reprogramming from human fibroblasts into functional neurons. Stem Cells Dev 2015;24(23):2746–55.

[109] Meyer S, Wörsdörfer P, Günther K, Thier M, Edenhofer F. Derivation of adult human fibroblasts and their direct conversion into expandable neural progenitor cells. J Vis Exp 2015;2015(101):e52831-e.

[110] Black JB, Adler AF, Wang H-G, D'Ippolito AM, Hutchinson HA, Reddy TE, et al. Targeted epigenetic remodeling of endogenous loci by CRISPR/Cas9-based transcriptional activators directly converts fibroblasts to neuronal cells. Cell Stem Cell 2016;19(3):406–14.

[111] Serguera C, Bemelmans AP. Gene therapy of the central nervous system: general considerations on viral vectors for gene transfer into the brain. Rev Neurol (Paris) 2014;170(12):727–38.

[112] Treutlein B, Lee QY, Camp JG, Mall M, Koh W, SAM S, et al. Dissecting direct reprogramming from fibroblast to neuron using single-cell RNA-seq. Nature 2016;534(7607):391–5.

[113] Han Y-C, Lim Y, Duffieldl MD, Li H, Liu J, Abdul Manaph NP, et al. Direct reprogramming of mouse fibroblasts to neural stem cells by small molecules. Stem Cells Int 2016;2015:1–11.

[114] Li X, Zuo X, Jing J, Ma Y, Wang J, Liu D, et al. Small-molecule-driven direct reprogramming of mouse fibroblasts into functional neurons. Cell Stem Cell 2015;17(2):195–203.
[115] Hu W, Qiu B, Guan W, Wang Q, Wang M, Li W, et al. Direct conversion of normal and Alzheimer's disease human fibroblasts into neuronal cells by small molecules. Cell Stem Cell 2015;17(2):204–12.
[116] Zheng J, Choi K-A, Kang PJ, Hyeon S, Kwon S, Moon J-H, et al. A combination of small molecules directly reprograms mouse fibroblasts into neural stem cells. Biochem Biophys Res Commun 2016;476(1):42–8.
[117] Tian E, Sun G, Sun G, Chao J, Ye P, Warden C, et al. Small-molecule-based lineage reprogramming creates functional astrocytes. Cell Rep 2016;16(3):781–92.
[118] Nawy T. Stem cells: fast track to neurons. Nat Methods 2015;12(10):915.
[119] Niu W, Zang T, Smith DK, Vue TY, Zou Y, Bachoo R, et al. SOX2 reprograms resident astrocytes into neural progenitors in the adult brain. Stem Cell Rep 2015;4(5):780–94.
[120] Guo Z, Zhang L, Wu Z, Chen Y, Wang F, Chen G. In vivo direct reprogramming of reactive glial cells into functional neurons after brain injury and in an Alzheimer's disease model. Cell Stem Cell 2014;14(2):188–202.
[121] Torper O, Pfisterer U, Wolf DA, Pereira M, Lau S, Jakobsson J, et al. Generation of induced neurons via direct conversion in vivo. Proc Natl Acad Sci 2013;110(17):7038–43.
[122] Torper O, Ottosson DR, Pereira M, Lau S, Cardoso T, Grealish S, et al. In vivo reprogramming of striatal NG2 glia into functional neurons that integrate into local host circuitry. Cell Rep 2015;12(3):474–81.
[123] Liu Y, Miao Q, Yuan J, Han S, Zhang P, Li S, et al. Ascl1 converts dorsal midbrain astrocytes into functional neurons in vivo. J Neurosci 2015;35(25):9336–55.
[124] Su Z, Niu W, Liu M-L, Zou Y, Zhang C-L. In vivo conversion of astrocytes to neurons in the injured adult spinal cord. Nat Commun 2014;5.
[125] Liu M-L, Zang T, Zhang C-L. Direct lineage reprogramming reveals disease-specific phenotypes of motor neurons from human ALS patients. Cell Rep 2016;14(1):115–28.
[126] Xue Y, Qian H, Hu J, Zhou B, Zhou Y, Hu X, et al. Sequential regulatory loops as key gatekeepers for neuronal reprogramming in human cells. Nat Neurosci 2016;19(6):807–15.
[127] Obokata H, Wakayama T, Sasai Y, Kojima K, Vacanti MP, Niwa H, et al. Stimulus-triggered fate conversion of somatic cells into pluripotency. Nature 2014;505(7485):641–7.
[128] Obokata H, Sasai Y, Niwa H, Kadota M, Andrabi M, Takata N, et al. Bidirectional developmental potential in reprogrammed cells with acquired pluripotency. Nature 2014;505(7485):676–80.
[129] De Los Angeles A, Ferrari F, Fujiwara Y, Mathieu R, Lee S, Lee S, et al. Failure to replicate the STAP cell phenomenon. Nature 2015;525(7570):E6–9.

CHAPTER

Design considerations when engineering neural tissue from stem cells

4

1 INTRODUCTION

This book has already addressed the structures and cells that make up the nervous system in Chapter 2. However, these tissues possess additional properties that must be considered when engineering replacement tissues by combining biomaterials, stem cells, and other components. To address these considerations, this chapter will examine in detail the chemical and mechanical properties of neural tissue and how we can replicate these properties using biomaterials and drug delivery systems as methods of controlling stem cell behavior. These topics will then be covered in detail by Chapters 5–7 with Chapters 5 and 6 discussing natural and synthetic biomaterials that have been applied to neural tissue engineering applications and Chapter 7 discussing different ways of delivering cues for controlling stem cell behavior.

All engineering projects have design constraints that must be addressed. Here we discuss the desired chemical and mechanical properties for engineering different types of neural tissues by examining the properties of healthy tissue as a starting point. The chapter ends by discussing clinical and commercial considerations when engineering tissues by combining biomaterial scaffolds with stem cells. One of the most obvious starting places for determining design criteria is the composition and structure of the extracellular matrix, which is the collection of biomolecules secreted by cells into the extracellular space [1]. The extracellular matrix supports the cells in native tissue, ensuring that they can perform their appropriate functions. One of the major roles of the extracellular matrix is to provide the necessary structures for proper cell function by presenting appropriate chemical and physical cues [2]. These chemical cues include the proteins and polysaccharides that make up the structural elements of the extracellular matrix where cells can bind, as well as the growth factors and other chemical molecules that act directly on cells to influence their behavior. Often the network of biomolecules can trap bioactive molecules like growth factors, resulting in the creation of reservoirs of these factors that act upon cells. The mechanical properties of the extracellular matrix also influence the behaviors of cells as well. These properties include the elastic modulus and topography associated with a material.

The properties of the extracellular matrix vary based on tissue type and this chapter will focus on how these properties are presented in the neural tissue [3]. As detailed in Chapter 2, neural tissue consists of multiple cell types organized

Engineering Neural Tissue from Stem Cells. http://dx.doi.org/10.1016/B978-0-12-811385-1.00004-2

in a complex fashion, which enables proper function. Thus, an optimal scaffold for engineering tissue from stem cells would enable the support of multiple cell types, allowing them to form three-dimensional (3D) structures that replicate the different scales of topography present in native tissue [4]. Finally, the resulting design should also consider the potential for clinical translation and commercial considerations, which will be discussed in Section 4. For example, the cells and materials used to construct engineered tissues should not trigger immune reactions after transplantation in vivo, and consideration should be given to how the necessary quantities of cells will be obtained. The cost of the total construct must also be considered along with its stability and necessary storage conditions as these factors will influence the potential of such engineered tissues to be marketed commercially [5].

2 CHEMICAL PROPERTIES OF NEURAL TISSUE

This section will cover the different molecules that work together to form the extracellular matrix along with the different types of chemical cues that control cell behavior relevant to neural tissue engineering. The extracellular matrix supports cells as they adhere to both the matrix and other adjacent cells and enables cells to communicate with each other [1,2]. Cells secrete the components that create the extracellular matrix. The composition of the extracellular matrix changes during the course of development, making its properties time dependent. In particular, the properties of the extracellular matrix are determined by the combination and concentrations of the different molecules present in the system. It takes an intricate combination of both positive and negative regulators of cell behavior to ensure proper tissue function.

Fig. 1 shows the different components of the extracellular matrix found in brain tissue. It is important to understand how the extracellular matrix functions in healthy neural tissue as it becomes altered in the case of neurodegenerative diseases, leading to dysfunction [6]. In general, two major classes of molecules make up the extracellular matrix: proteins and polysaccharides. Proteins are polymer chains of amino acids that perform a wide range of roles in the body. These proteins have different levels of structure and often a fully functional protein can be made up of multiple subunits. For example, many growth factors consist of two subunits and are referred to as dimers. The proteins of the extracellular matrix tend to be fibrous in nature, as opposed to globular, because they act to ensure that cells have proper structural support. Polysaccharides, as their name implies, consist of polymers built from monosaccharide subunits, making them long sugar chains. The diversity of monomers for both types of molecules means the resulting polymers differ widely in their structure and function, allowing them to vary widely in their properties and enabling them to perform unique functions. More detailed information about the general properties of the extracellular matrix can be found in the classic textbooks written on cell biology listed in the resource section given at the end of this chapter.

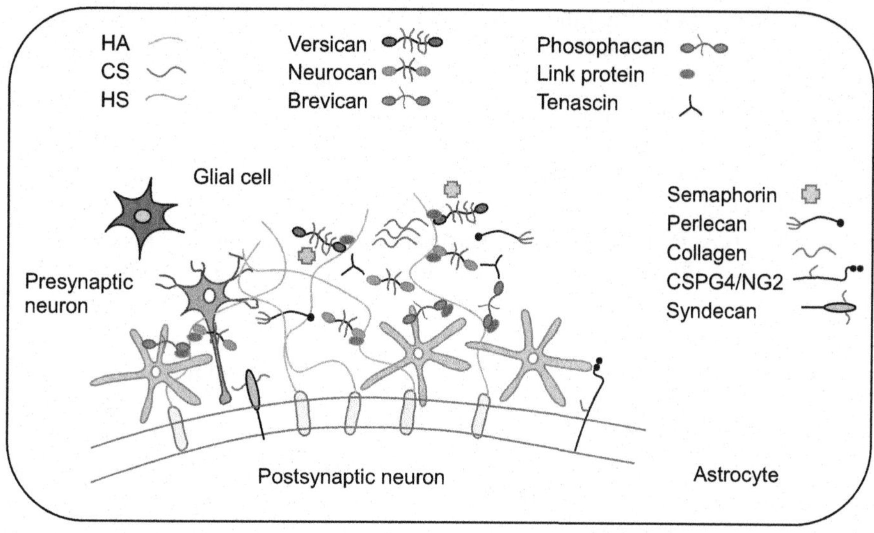

FIG. 1

The molecules found in the extracellular matrix of the brain. These molecules include heparan sulfate proteoglycans (HS), chondroitin sulfate proteoglycans (CS), hyaluronic acid (HA), collagen, and other glycoproteins.

Reprinted with permission from Sethi MK, Zaia J. Extracellular matrix proteomics in schizophrenia and Alzheimer's disease. Anal Bioanal Chem 2016:1–16.

2.1 PROTEINS OF THE EXTRACELLULAR MATRIX AND THEIR FUNCTION

This section will define the major classes of proteins found in the extracellular matrix and discuss their chemical properties. Understanding how their chemical properties contribute to their ability to support cells and their function will enable the development of successful replacements for engineering neural tissue from stem cells. Chapter 5 will explore how such proteins can function as 3D biomaterial scaffolds that support stem cell culture and differentiation into neural phenotypes. The central nervous system contains three distinct structures that arise from the extracellular matrix: the interstitial spaces, the basement membrane, and the perineuronal nets. Briefly summarized, the interstitial space is the small regions found in between the cells that make up a particular organ while the basement membrane serves as a border between the epithelium and tissues. Finally, perineuronal nets are specialized regions of the extracellular matrix that ensure proper transmission of the signals relayed by neurons. These regions will be further discussed when appropriate in the following sections. Later in this chapter, the differences in extracellular matrix composition will be compared between the different organs of the central nervous system and the peripheral nervous system.

2.1.1 Collagen

The protein collagen derives its name from the Greek words meaning "glue producing." This protein connects cells together, creating structural integrity in the extracellular matrix through the body. Accordingly, it is the most abundant protein found in the extracellular matrix, as well as in the human body, due to this important physiological role. The different types of collagen can be classified broadly into fibrillar and non-fibrillar forms. These proteins can also be further divided into more specific classifications based on types such as Types I, II, III, and so on. Over 28 different types of collagen have been identified, but this chapter will only focus on Types I, III, and IV, which are found in nerve tissue. Type I and III collagens are fibrillar in nature while Type IV is non-fibrillar. Peripheral nerve tissue contains mainly Type I and III collagens [7] although other collagens are present in smaller quantities where they contribute to cell signaling [8]. Type IV collagen plays a unique role in the body, contributing a structure known as the basement membrane. The basement membrane separates the different layers of tissues found in the body, including the blood-brain barrier [9]. The differences between the types of collagen often result from how the cell processes this protein. Cells initially produce a precursor protein named procollagen, which is then processed into functional collagen by enzymes called proteases [10]. Additionally, denatured collagen is referred to as gelatin. Despite being denatured, gelatin does retain bioactive properties [11].

2.1.2 Fibronectin

The protein fibronectin possesses a high molecular weight, consisting of a dimer where each protein subunit weighs 220 kDa. It can be found in two forms in vivo. The first form is produced by the liver and it circulates in the blood, while the other insoluble form secreted by fibroblasts exists to stabilize the extracellular matrix. Fibronectin also contains polysaccharide chains as part of its structure in addition to its protein subunits. The large size of fibronectin enables it to act as the master organizer of the extracellular matrix [12]. Accordingly, it performs several functions, including allowing cells to bind to it through proteins called integrins and retaining bioactive factors inside the extracellular matrix. Patterns of fibronectin expression change over the course of tissue development; fibronectin tends to be highly expressed during nerve development, followed by lower expression in mature nerve tissue [13]. Fibronectin regulates stem cell behavior and differentiation, and it has been used as a two-dimensional (2D) substrate for neural stem cell culture [14]. Fibronectin's ability to bind integrins can also influence the resulting mechanical properties of the cells inside the matrix, and the ability of the matrix to bind bioactive factors and sequester them for later use, making this material of interest when engineering neural tissue.

2.1.3 Laminin

Laminin is a protein that shares several properties with fibronectin. For example, it also has a high molecular weight. The laminin protein consists of three subunits, including α, β, and γ chains as shown in Fig. 2. Laminins are also considered to be glycoproteins as they are functionalized with oligosaccharides [3]. In the extracellular matrix, laminin can bind other laminin molecules as well as other proteins like collagen, which helps to reinforce the extracellular matrix structure. Cells can also bind

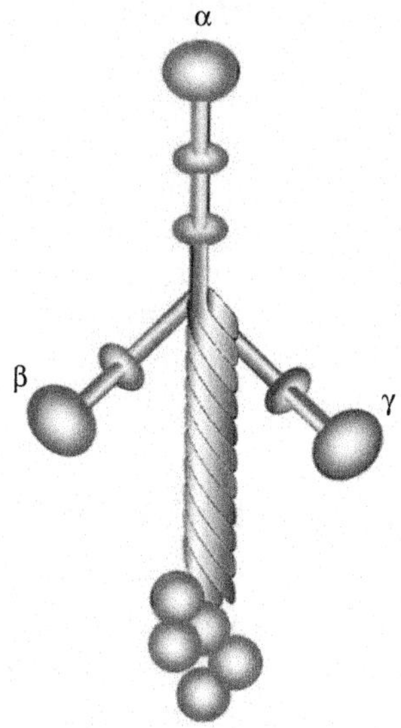

FIG. 2

The molecular structure of laminin. It consists of three subunits named α, β, and γ.

Reprinted with permission from Plantman S. Proregenerative properties of ECM molecules. BioMed Res Int 2013;2013.

to laminin through the integrin receptors they express on in their cell membranes [15]. These properties make laminins attractive for use as cell culture substrates for both pluripotent stem cells and neural cells [16,17]. Protocols for differentiating stem cells into mature phenotypes will often require culture on 2D laminin substrates. Laminin also plays a key role in the development of the nervous system as axons tend to migrate on this protein, making it attractive for regenerative medicine applications [18]. Finally, laminin is found in the neural stem cell niche, where it contributes its growth factor binding properties to specialized structures known as fractones [19].

2.1.4 Elastin

Unlike the proteins discussed previously in this section, elastins, as their name implies, are highly elastic in nature, and function by allowing tissues to stretch [20]. Their presence also allows nerve tissue to respond to deformation without breaking [21]. Elastin is derived from the precursor protein tropoelastin and it can be spliced into 11 different isoforms, which are variants of the same protein [22]. It forms cross-links between fibrillary proteins, generating elasticity in a tissue. When compared to other tissues like blood vessels, nerve tissue contains relatively smaller amounts of elastin [23]. The combination and concentration of these proteins determines the mechanical properties of the different types of tissue found in the nervous system. These properties and the effects they have on cell behavior will be discussed later in this chapter.

2.1.5 Glycoproteins found in the extracellular matrix of the nervous system

In addition to above proteins, several glycoproteins also make up the extracellular matrix of the nervous system. For example, the radial glia cells express tenascin-c, which is an important component of the neural stem cell niche [24]. Other glycoproteins that play important roles in the nervous system include versican, neurocan, and brevican. Versican is expressed in the neural crest during development [25]. Neurocan and brevican plays important roles in ensuring proper migration of neural stem cells in the brain [26]. These components can be incorporated into scaffolds for promoting desired behavior of the stem cells seeded inside.

2.2 POLYSACCHARIDES FOUND IN THE EXTRACELLULAR MATRIX AND THEIR FUNCTION

Polysaccharides remain relatively understudied in comparison to the wide body of knowledge associated with protein structure and function in the extracellular matrix. These molecules play significant biological roles in vivo. In the extracellular matrix, polysaccharides, chains of sugar monomers covalently linked together, can function as independent molecules such as hyaluronic acid and in combination with proteins as part of complexes called proteoglycans [20]. Proteoglycans consist of a protein core functionalized with polysaccharides, increasing their ability to regulate cellular functions. The term glycan is a synonym for polysaccharide. These biomolecules tend to vary in composition and molecular weight as their synthesis is more heterogeneous compared to how proteins are made. The process where these glycans are attached to a protein is referred to as glycosylation, which refers to the chemical bond that links these molecules together [27]. The type of bond is determined based on where the sugar molecule is added to the protein. N-linked glycosylation occurs when the sugar reacts with an amino group containing nitrogen, while O-linked glycosylation occurs when the sugar reacts with hydroxyl groups. Proteoglycans can also have varying degrees of sulfation, characterized by sites that are modified to contain the side chain: R–S(=O)2–OH where R is the atom on the molecule being modified [28]. Proteoglycans also play important roles in axonal guidance, making them a potential target for incorporation into biomaterial scaffolds [29]. They can also influence stem cell behavior [30] and have been implicated in neurological diseases like Alzheimer's [31].

2.2.1 Hyaluronan/hyaluronic acid

Several tissues in the body contain large amounts of hyaluronic acid, which is also referred to as hyaluronan. While the highest levels of hyaluronic acid are found in the skin, and accordingly in different types of skin creams, the brain and nervous system also contain significant quantities of this polysaccharide [32]. The cell membrane serves as the site of hyaluronic acid synthesis due to its size, which can cover a wide range of sizes. Low molecular weight hyaluronic acid has a molecular weight of <25 kDa while the high molecular weight version must have a mass of >400 kDa

[33]. The size of this polysaccharide determines its role in the extracellular matrix, making it a versatile biosignal. High molecular weight hyaluronic acid plays a structural role in the extracellular matrix and it regulates the amount of water content present in a tissue. Intriguingly, this version of hyaluronic acid has neuroprotective properties, and it reduces the amount of glial scarring post-injury by modulating the behavior of astrocytes [34]. Low molecular weight hyaluronic acid serves as a signaling molecule. CD44+, a transmembrane protein, serves as the main cell surface receptor that can bind hyaluronic acid [35]. If this protein sounds familiar, that is because it was previously discussed as a cell surface marker expressed by mesenchymal stem cells in Chapter 3. Interestingly, while the role of hyaluronic acid has been well characterized in the hematopoietic stem cell niche, its role in the neural stem cell niche remains understudied despite its significant presence there [33]. It does play a role in modulating the behavior of neural stem cells and more details on these studies will be given in Chapter 5. Overall, hyaluronic acid serves a major component of the extracellular matrix present in neural tissue and it possesses many attractive properties that are beneficial when engineering neural tissue from stem cells.

2.2.2 Chondroitin sulfate

The polysaccharide chondroitin sulfate consists of alternating subunits of the sugars *N*-acetyl galactosamine and glucuronic acid that form a chain [28]. The length and structure of this chain may vary, creating a diverse range of molecules that are categorized as chondroitin sulfate. This proteoglycan is often associated with the nervous system in the context of injuries, and where these molecules inhibit regeneration they are referred to as chondroitin sulfate proteoglycans (CSPGs) [36]. These CSPGs contribute to the formation of the glial scar post trauma in the central nervous system. Many strategies for promoting regeneration post-injury focus on how to counteract the effects of these molecules on neural regeneration. These molecules also bind other growth factors, making them of interest for applications in drug delivery, as detailed in a recent review [37]. Chondroitin sulfates also play key roles in regulating development, making them of interest to the stem cell community [38]. These molecules provide structural stability to the perineuronal nets, which are regions of the extracellular matrix found in the brain that stabilize synapses—the junction between neurons where signaling occurs [39].

2.2.3 Heparin and heparan sulfate

Heparin and heparan sulfate are both polysaccharides that perform important roles in the central nervous system. Fig. 3 shows the differences in structures between the two molecules as heparin tends to have higher degree of sulfation compared to heparin sulfate. While heparin acts as an anticoagulant in the bloodstream after being released by mast cells, it is not usually found in tissues [40]. The form of heparin that is secreted into the blood is not attached to a protein. However, heparin can bind growth factors due to its negatively charged nature, and accordingly, drug delivery systems take advantage of this property to deliver such factors in a biomimetic fashion [41]. It can also be used directly as a biomaterial scaffold. Heparan sulfate, on

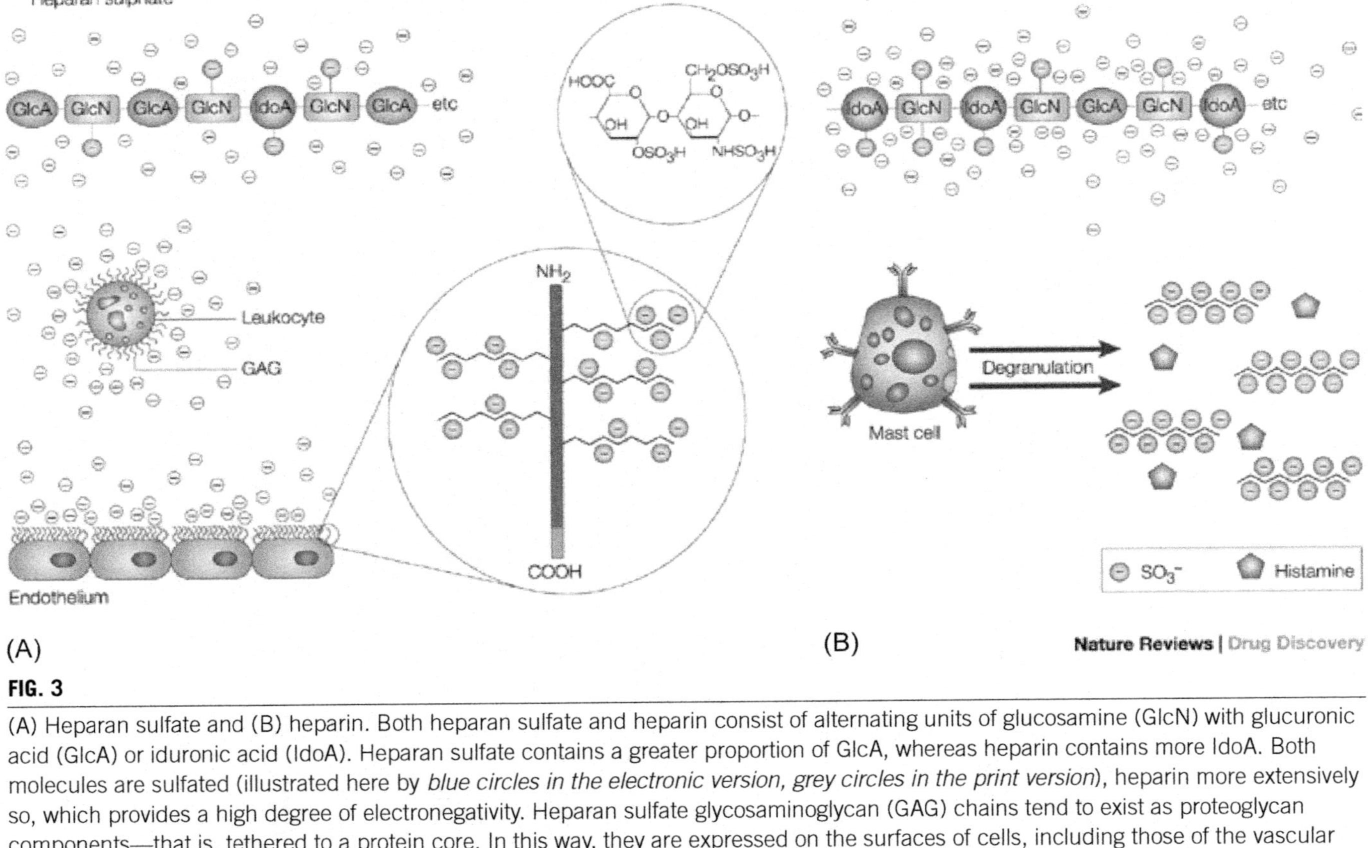

FIG. 3

(A) Heparan sulfate and (B) heparin. Both heparan sulfate and heparin consist of alternating units of glucosamine (GlcN) with glucuronic acid (GlcA) or iduronic acid (IdoA). Heparan sulfate contains a greater proportion of GlcA, whereas heparin contains more IdoA. Both molecules are sulfated (illustrated here by *blue circles in the electronic version, grey circles in the print version*), heparin more extensively so, which provides a high degree of electronegativity. Heparan sulfate glycosaminoglycan (GAG) chains tend to exist as proteoglycan components—that is, tethered to a protein core. In this way, they are expressed on the surfaces of cells, including those of the vascular endothelium and of circulating leukocytes, which provides a general net negative charge to these surfaces (depicted by *white circles in the print and electronic versions*). Heparin, by contrast, is co-released with histamine from degranulating mast cells, and can dissociate from its protein core to exist as free GAG chains.

Reprinted with permission from Lever R, Page CP. Novel drug development opportunities for heparin. Nat Rev Drug Discov 2002;1(2):140–8.

the other hand, is commonly found in the extracellular matrix where it modifies proteins, making them proteoglycans. Heparan sulfate, as its name implies, is also highly sulfated and negatively charged. These proteoglycans functionalized with heparan sulfate play major roles in the nervous system [42]. The proteoglycans' roles include promoting the differentiation of stem cells in the developing nervous system to form mature neurons, guiding axons into place, and the formation and stabilization of synapses. Heparan sulfate also regulates embryonic stem cell behavior [43]. Furthermore, heparan sulfate serves as one of the major components of the fractones, novel regions of extracellular matrix found in the neural stem cell niche that regulate growth factor levels [44].

2.2.4 Keratan sulfate/keratosulfate

Like chondroitin sulfate and heparan sulfate, keratan sulfate is classified as a proteoglycan. While bone and cartilage express high levels of keratan sulfate, it is not highly expressed in the mature nervous system except in the cornea [45]. Fig. 4 shows the two major classes of keratan sulfate, which are classified based on the tissues they were isolated from—either the cornea or cartilage. Like chondroitin sulfates, keratan sulfates also regulate the developing nervous system [46]. In particular, keratan sulfates guide axons located in the thalamus, as opposed to axons of the cortex. The expression of keratan sulfates varies throughout the brain and potential

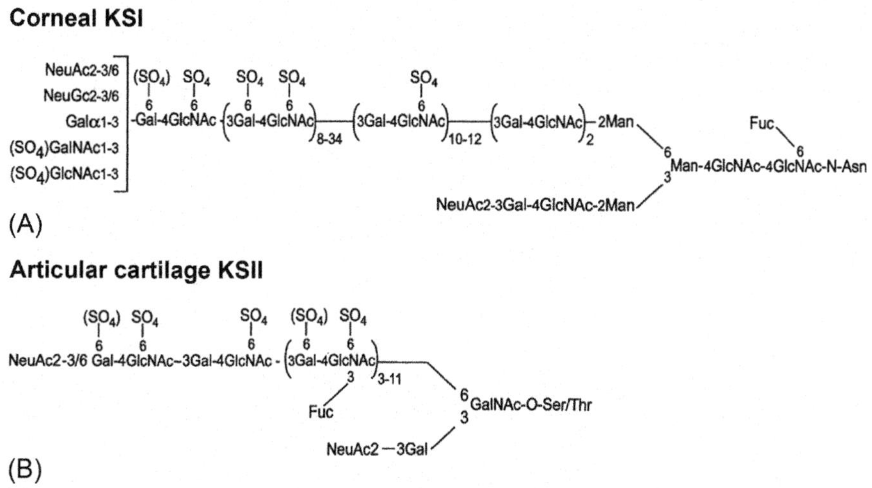

Corneal KSI

(A)

Articular cartilage KSII

(B)

FIG. 4

Summary of structural features of keratan sulfate molecules. (A) Corneal keratan sulfate I. (B) Articular cartilage keratan sulfate II. Numbers to the lower right of the large parenthesis show the approximate number of *N*-acetyl lactosamine monomers in each domain. Numbers separated by a forward slash present optional attachment locations. Sulfates in parentheses indicates partial or incomplete sulfation of monomers at this site.

Reprinted with permission from Funderburgh JL. MINI REVIEW Keratan sulfate: structure, biosynthesis, and function. Glycobiology 2000;10(10):951–8.

tissue engineering strategies could exploit this property for producing specific neuronal subtypes from stem cells. Keratan sulfates are also expressed post-injury in the central nervous system and contribute to glial scarring [47]. Thus, targeting the expression of these molecules may provide a way to create a more permissive environment for regenerative after traumatic injuries.

2.3 COMPOSITION OF THE EXTRACELLULAR MATRIX OF THE BRAIN AND THE SPINAL CORD

The composition of the extracellular matrix varies depending on the region of the nervous system. These sections will highlight these differences to provide insight into how to design materials for replicating these regions with the goal of promoting stem cell differentiation into the desired phenotypes. While the brain and spinal cord differ in their extracellular matrix composition, perineuronal nets, which are found in both locations, do exhibit similar compositions of their specialized extracellular matrix [48]. A 2013 Biomaterials paper compared the composition of decellularized extracellular matrix isolated from the brain with the matrix isolated from the spinal cord [49]. They found higher levels of collagen content in the spinal cord in comparison to the brain. However, both tissues contained similar levels of glycosaminoglycans. They also found that the storage modulus (the portion of a material that responds elastically under stress) was higher for the extracellular matrix isolated from the spinal cord than the extracellular matrix isolated from brain tissue. They attributed these differences to the levels of collagen content in the tissues. Both of these isolated matrices supported neural cell culture and neurite outgrowth, indicating the specialized nature of different regions of extracellular matrix.

Further work has explored how the composition of the extracellular matrix changes in response to injury. Much of this work has been done in the context of spinal cord injury due to the involvement of CSPGs, which inhibit regeneration. Fig. 5 shows how the extracellular matrix of the spinal cord changes in response to injury, and details how this response develops over time as the spinal cord injury progresses through the acute and chronic stages [50]. In response to injury, cells begin to secrete matrix metalloproteases that degrade the proteins, such as collagen, laminin, and fibronectin, present in the spinal cord. These components tend to promote regeneration, so their destruction starts the process of generating an inhibitory environment. This process is coupled with the remaining cells, usually reactive astrocytes, upregulating expression of CSPGs, which also prevent regeneration. In the acute stage of injury, damaged blood vessels contribute to these effects. In the chronic stage of injury, the blood vessels and extracellular matrix stabilize along with the injury site, which now consists of a glial scar [51]. Proteomic analysis has also demonstrated that in certain neurological disorders such as Alzheimer's disease and schizophrenia, extracellular matrix also changes from its normal composition into a less supportive environment for promoting cell survival and function [6]. Thus, the use of factors shown to promote neural cell survival can help supplement the survival of cells being transplanted into the injured nervous system.

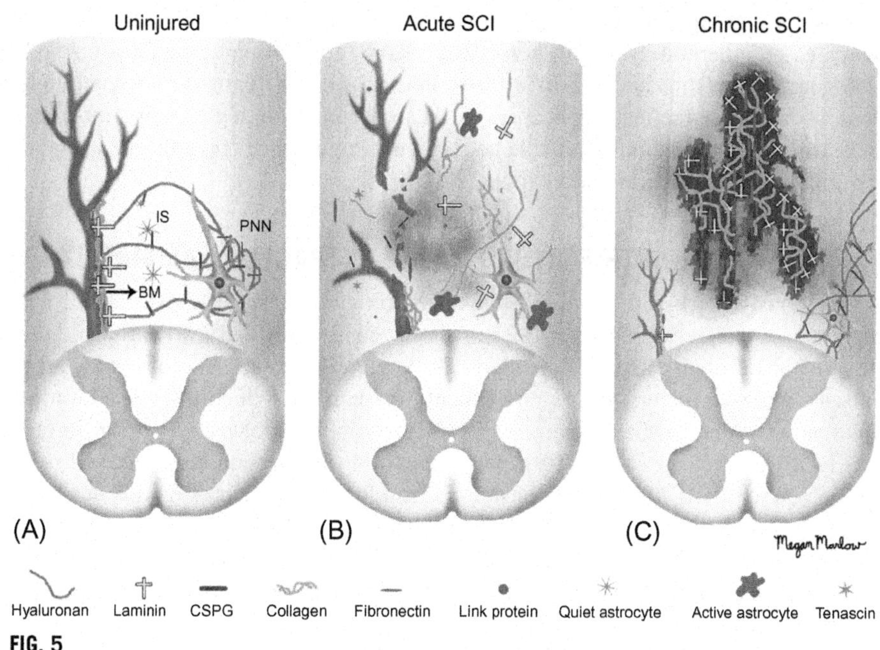

FIG. 5

Cartoon representing the location and association of extracellular matrix in healthy uninjured spinal cord and at the acute and chronic stage of spinal cord injury (SCI) in the adult rat. (A) Uninjured spinal cord in which collagen, fibronectin, and laminin are in close association with each other in the perineuronal net (PNN), interstitial space (IS), and the basal membranes (BM) of blood vessels. (B) In acute SCI blood vessels are damaged and extracellular matrix (ECM) is broken down and is in disarray. Reactive astrocytes are present and there is significant neuron and axon damage. (C) In chronic SCI tissue damage is extensive. Some blood vessels remain or new ones have formed. Collagen, fibronectin, and laminin are secreted from cells in and near the injury site. Scar tissue that has formed at the injury site is especially rich in proteoglycans, which are inhibitory to axon growth. Laminin can be found surrounding the injury inside of the astrocytic scar.

Reprinted with permission from Haggarty, A., Marlow, M., Oudega, M. (2016). Extracellular matrix components as therapeutics for spinal cord injury. Neuroscience Letters.

2.4 COMPOSITION OF THE EXTRACELLULAR MATRIX IN THE PERIPHERAL NERVOUS SYSTEM

As discussed previously in this book, the peripheral nervous system possesses a much higher capacity for regeneration in comparison to the central nervous system [52]. This regenerative capacity reflects the presence of Schwann cells, which respond to injury by secreting factors that stimulate regeneration. Although CSPGs still have an effect on regeneration in the peripheral nervous system, it is not as severe as in the central nervous system [53]. The peripheral nervous system contains mainly collagen along with the other aforementioned components of the extracellular matrix

[54]. Several groups have evaluated the use of nerve guidance conduits for repairing damaged peripheral nerves. Different studies have found filling these conduits with collagen, fibronectin, and fibrin (a blood-derived protein) enhances regeneration, demonstrating a practical application of these proteins for regenerative medicine [55]. In fact, commercially available collagen nerve guidance conduits are available for clinical use in repairing nerve injuries [56].

2.5 ROLE OF GROWTH FACTORS IN THE EXTRACELLULAR MATRIX

As previously discussed, the extracellular matrix can entrap bioactive factors that regulate cell behaviors. Many of these factors promote cell survival and proliferation and accordingly they are considered to be growth factors. Neurotrophins are a subset of growth factors that act specifically on neurons, which often used during applications in neural tissue engineering. Neurotrophins include nerve growth factor (NGF), neurotrophin-3 (NT-3), neurotrophin-4 (NT-4), and brain-derived neurotrophic factor (BDNF). Improper internalization of these neurotrophins has been implicated in developmental neurological disorders, as these proteins mediate the creation of proper neural connections [57]. While NGF is the best characterized neurotrophin proteins, it is important to select the appropriate factor depending on the desired effect.

Dr. Rita Levi-Montalcini and Dr. Stanley Cohen, who were working at Washington University in St. Louis, were the first to identify the presence of growth factors and the growth factor they initially discovered was NGF [58]. NGF targets specific populations of neurons present in both the central and peripheral nervous system. In the central nervous system, NGF acts mainly on cholinergic neurons while in the peripheral nervous system it targets sensory neurons. The second neurotrophic factor identified was BDNF, which shares a high degree of similarity with NGF [59]. It promotes the survival of the neurons found in dorsal root ganglia. Other members of the neurotrophin family include NT-3 and NT-4 [60]. NT-3 promotes neuronal survival as well as neurogenesis. It targets a wider range of neuronal subtypes compared to NT-4. NT-4 is comparatively understudied to the other members of the neurotrophin family. Neurotrophins act by binding a family of receptor tyrosine kinases called Trk receptors, which activate intracellular signaling cascades that result in cell survival, proliferation, and differentiation [61]. They also bind to another receptor called p75, which also regulates these functions [62]. Thus, these growth factors act on different cell populations based on which of these aforementioned receptors they express.

In addition to neurotrophins, other growth factors have been identified that influence the behavior of both stem cells and cells of the nervous systems. These factors include basic fibroblast growth factor (bFGF), epidermal growth factor (EGF), platelet-derived growth factor (PDGF), glial-derived neurotrophic factor (GDNF), sonic hedgehog (Shh), stromal-derived factor 1 (SDF1), and a family of proteins called the ephrins. bFGF and EGF serve as general growth factors that promote cell survival and proliferation and they also promote neurite outgrowth in neuronal cells to different degrees [63]. These factors are often included when delivering cells into the damaged nervous system to increase the rate of cell survival.

PDGF plays an important role in regulating the maturation of oligodendrocytes and during neurogenesis, making it useful for promoting stem cell differentiation into these phenotypes [64]. Shh also plays important roles in the development of multiple tissues, including the patterning of the nervous system [65]. It regulates neural progenitor proliferation during adult neurogenesis [66] and it also directs the formation of the neural crest [67]. Recent studies have indicated that more potent versions of Shh can be engineered through multiplexing these proteins. These more potent versions better replicate the biological effects of Shh in vitro and in vivo compared to Shh produced using traditional recombinant protein expression methods [68]. The cells in the neural stem cell niche described in Chapter 2 secrete SDF1 as a growth factor for maintaining neural progenitors [69], making it an interesting target for therapeutic applications that use neural stem cells. Finally, a family of proteins known as ephrins perform important functions by binding to receptor tyrosine kinases called eph receptors [70]. Intriguingly, both ephrins and eph receptors are membrane bound proteins, which makes them difficult to study in comparison with other growth factors. Similar to other factors described, these proteins regulate axonal guidance in developing the nervous system, making them of interest when engineering the neural tissue [71]. Recent work has demonstrated that delivering a combination of multiplexed ephrin and SDF1 can induce neurogenesis in non-neurogenic regions of the brain, mimicking the neural stem cell niche [72]. Overall, a number of proteins can be used for applications in neural tissue engineering.

2.6 THE ROLE OF SMALL MOLECULES IN NEURONAL DEVELOPMENT

In addition to growth factors, small molecules also play important roles in the development of the nervous system and in differentiating stem cells into mature neural phenotypes. One of the most commonly used small molecules is retinoic acid, a Vitamin A derivative, which plays a major role in patterning the nervous system during development [73]. It activates the Hox genes, and its activity results in the development of the hindbrain. It is a potent morphogen for differentiating pluripotent stem cells into neurons [74]. Another small molecule of interest when engineering neural tissue is purmorphamine. It acts as a Sonic Hedgehog agonist, and it promotes pluripotent stem cell differentiation into motor neurons [75–78]. Finally, a small molecule called guggulsterone can efficiently generate dopaminergic neurons from human pluripotent stem cells [79,80]. All of these small molecules can be incorporated into biomaterial scaffolds for applications in engineering neural tissue and serve as alternatives to using growth factors to influence stem cell behavior.

3 MECHANICAL PROPERTIES OF NEURAL TISSUE

The bioactive molecules detailed in Section 2 also contribute to the mechanical properties of the tissues found in the central and peripheral nervous system. While traditionally more focus has been placed on how chemicals and biological molecules can influence cell behavior, attention has shifted in recent years to focus on how the mechanical properties of scaffolds can be used to control cells, especially in the context of stem cell differentiation [81]. This section will describe how the physical topography and associated mechanical properties of tissue ensure the proper function of the nervous system and how these mechanical properties can be exploited when engineering new substrates for directing stem cell differentiation into neural phenotypes.

3.1 THE PHYSICAL TOPOGRAPHY OF THE EXTRACELLULAR MATRIX AND ITS EFFECT ON CELLULAR BEHAVIOR

The fibrillar nature of the proteins and polysaccharides of the extracellular matrix found in the nervous systems results in both nanoscale and microscale topography, which influences cell function. The physical and mechanical properties of these structures can be replicated to determine their effect independent of the chemistry of the biological molecules found in vivo. The nervous system has a uniquely defined architecture that enable the neurons to be aligned, allowing them to perform their role in transmitting information effective [82]. One popular method for replicating these topographical effects uses a process called electrospinning to produce nano and microscale topographies that replicate these features found in nature [83–85]. Substrate topography significant impacts how stem cells behave both in vitro and in vivo. In particular, using aligned nanofibers enhances embryonic stem cell differentiation into neurons and the neurites extended by these neurons are longer on aligned scaffolds [86,87]. Such topographical cues can be incorporated into biomaterial scaffolds, where they enhance peripheral nerve regeneration [88]. Yang and Leong reviewed different methods for generating nanoscale topography using a variety of methods, including examples of how these topographies can influence the behavior of stem cells [89]. Larger features in the microscale range also influence stem cell behavior. For example, aligned microfiber grooves promote neuronal differentiation of human embryonic stem cells [90]. My own research group also showed that a novel microfiber architecture consisting of biaxial aligned scaffolds enhanced neurite extension from neurons derived from human-induced pluripotent stem cells while guiding these extensions along the fibers, demonstrating how contact guidance can be used to pattern cells into the desired structures [91]. These effects are often synergistic when combined with chemical and genetic cues, making them important considerations when engineering neural tissue from stem cells [91–96]. While the studies covered in this section are not exhaustive, they serve as important examples of how both nanoscale and microscale topography influence stem cell behavior and differentiation.

3.2 THE MECHANICAL PROPERTIES OF NEURAL TISSUE AND HOW THEY RELATE TO STEM CELL BEHAVIOR

As discussed previously, the extracellular matrix provides structures for cells to attach, usually through integrins—proteins expressed on the cell surface [97].

These interactions also play important roles in how stem cells behave, and they can be exploited for use when engineering tissues [98]. The surface receptors rely on both chemical and physical signals to transmit information to the cell. Mechanical cues are sensed through a process called mechanotransduction that involves integrins detecting the physical properties of their substrate [99]. A ground-breaking study conducted by Dr. Adam Engler (now at the University of California-San Diego) while he was working in Dr. Discher's lab at the University of Pennsylvania demonstrated how substrate stiffness affected stem cell differentiation [100]. This work showed that changing the elastic modulus of a substrate can determine the mature fate of a stem cell. The elastic modulus of a material is defined by its stress-strain curve and it reflects the ability of a material to resist being deformed by an outside force. Thus, a higher modulus indicates a stiffer substrate. They found that substrates that ranged from 0.1 to 1 kPa promoted mesenchymal stem cells to differentiate into neuron-like structures as shown in Fig. 6. This range of stiffness was similar to the mechanical properties of in vivo brain tissue. Further work from the Schaffer and Kumar groups expanded on this idea that mechanical properties could modulate the behavior of neural stem cells [101]. Their work showed that increasing the stiffness of the substrate would inhibit the migration of neural stem cells [102]. Other groups showed that higher substrate stiffness presented by alginate gels inhibited neuronal outgrowth from primary neurons [103]. Other groups have found that $G^* = 53$ kPa is the optimal range of stiffness for promoting the proliferation of Schwann cells [104].

In addition to the effects observed for controlling neural cell behavior, these physical cues influence the progress of reprogramming adult mature cells into induced pluripotent stem cells [105]. Other work has indicated that white matter tends to be less stiff than gray matter [106]. The mechanical properties of neural tissue also change during development, influencing both stem cell migration and the formation of the nervous system. It has also been hypothesized that these forces play critical roles in the formation of the nervous system [107]. For example, maintaining proper levels of tension ensures that neurons can function effectively. Overall, these studies indicate the importance of considering the mechanical properties of the desired tissue when developing scaffolds to mimic these functions.

4 IMPLICATIONS FOR DESIGN OF BIOMATERIAL SCAFFOLDS

The previous sections covered the desired chemical and mechanical properties found in healthy neural tissue that can be implemented using tissue engineering. However, additional considerations exist when designing such engineered tissues if they are to be implemented for clinical practice. One of the major considerations when designing such scaffolds is ensuring that they are biocompatible with the human body. Thus, a material should not be toxic to human cells nor should it cause other damage to the body—either directly or indirectly through activation of the immune system. Resources for learning more about how the body reacts to implanted materials can be

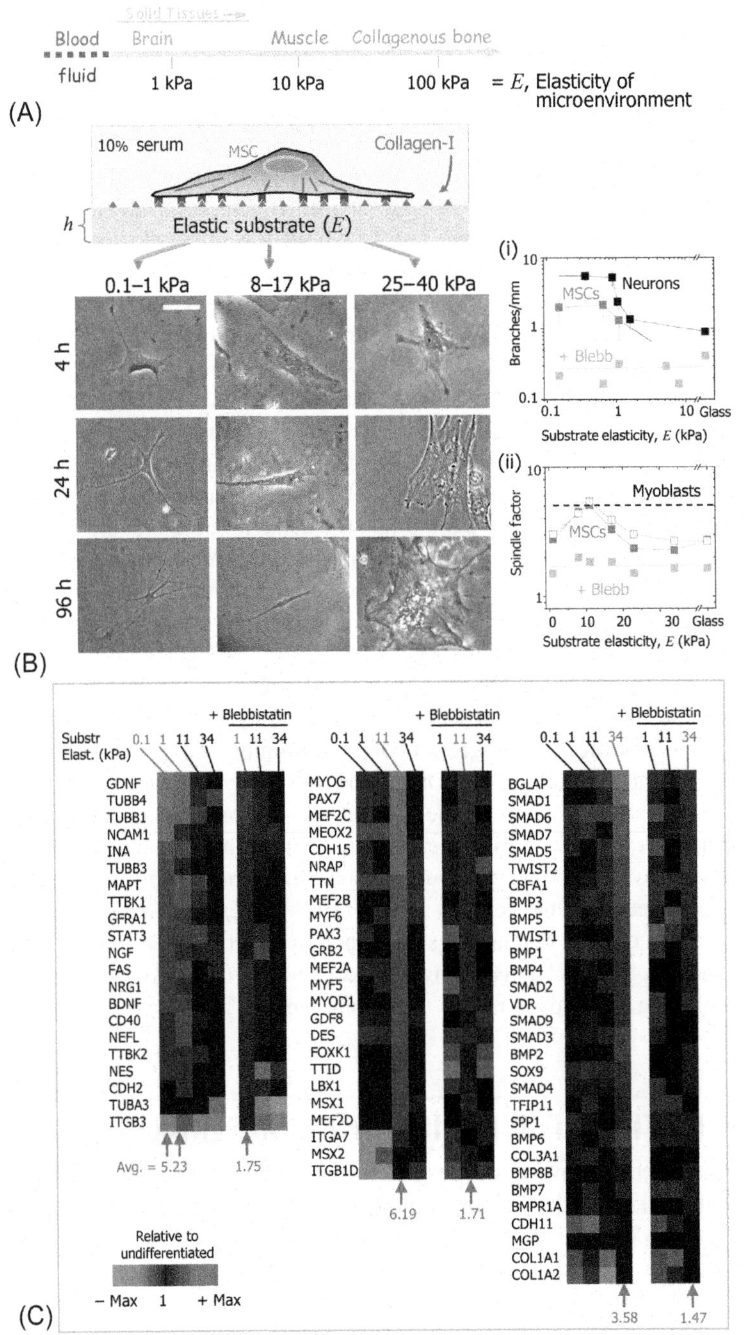

FIG. 6

See the legend on opposite page.

found at the end of this chapter. The same consideration also applies when sourcing cells to use for neural tissue engineering. The overall stability of the construct also should be considered—both in terms of the materials used and the cells necessary. Both maintaining the bioactivity of growth factors and keeping cells alive often limit the shelf life of such tissue engineered products. Another consideration is how to deliver such tissues in vivo. Surgeons often prefer injectable formulations as this delivery method makes it easy to bypass to the blood–brain barrier and the brain-spinal cord barrier [108].

One emerging area of research focuses on the use of 3D printers to generate tissues on demand. Companies working in this area include Organovo and Aspect Biosystems, with Organovo having successfully printed neural tissue using its system. The use of 3D printers also lends itself well to producing such tissues in a high throughput manner, which is another important consideration for clinical and commercial applications. Finally, the cost of these constructs should be minimized whenever possible. Recent clinical trials using stem cells to treat spinal cord injury have shown improved function posttreatment, but the overall cost of the procedure has made it hard to justify, in terms of making this stem cell treatment the standard of care for patients suffering from such spinal injuries. In conclusion, a wide variety of issues must be considered when designing such engineered tissues. The next two chapters will discuss relevant examples of using biomaterial scaffolds combined with stem cells to engineer neural tissue in the context of these design considerations.

FIG. 6

Tissue elasticity and differentiation of naive mesenchymal stem cells (MSCs). (A) Solid tissues exhibit a range of stiffness, as measured by the elastic modulus, E. (B) The in vitro gel system allows for control of E through cross-linking, control of cell adhesion by covalent attachment of collagen-I, and control of thickness, h. Naive MSCs are initially small and round but develop increasingly branched, spindle, or polygonal shapes when grown on matrices, respectively, in the range typical of ~Ebrain (0.1–1 kPa), ~Emuscle (8–17 kPa), or stiff cross-linked–collagen matrices (25–40 kPa). Scale bar is 20 μm. Inset graphs quantify the morphological changes (mean ± SEM) versus stiffness, E: shown are (i) cell branching per length of primary mouse neurons, MSCs, and blebbistatin-treated MSCs and (ii) spindle morphology of MSCs, blebbistatin-treated MSCs, and mitomycin-C–treated MSCs *(open squares)* compared to C2C12 myoblasts *(dashed line)*. (C) Microarray profiling of MSC transcripts in cells cultured on 0.1, 1, 11, or 34 kPa matrices with or without blebbistatin treatment. Results are normalized to actin levels and then normalized again to expression in naive MSCs, yielding the fold increase at the bottom of each array. Neurogenic markers (left) are clearly highest on 0.1–1 kPa gels, while myogenic markers (center) are highest on 11 kPa gels and osteogenic markers (right) are highest on 34 kPa gels. Blebbistatin blocks such specification (<twofold different from naive MSCs).

Reprinted with permission from Engler AJ, Sen S, Sweeney HL, Discher DE. Matrix elasticity directs stem cell lineage specification. Cell 2006;126(4):677–89.

ADDITIONAL RESOURCES

CELL BIOLOGY TEXTBOOKS FOR FURTHER READING ABOUT THE EXTRACELLULAR MATRIX

Lodish, H., Berk, A., Kaiser, C., Krieger, M., Bretscher, A., Ploegh, H., Amon, A., and Martin, K. Molecular Cell Biology, 8th Edition. New York: W.H. Freedman. 2016.

Alberts, B., Johnson, A., Lewis, J., Morgan, D., Raff, M., Roberts, K. Walter, P. Molecular Biology of the Cell. 6th edition. New York: Garland Science. 2014.

http://garlandscience.com/product/isbn/9780815344322

Alberts, B., Bray, D., Hopkin, K., Johnson, A., Lewis, J., Raff, M., Roberts, K., Walter, P. Essential Cell Biology. 4th edition. New York: Garland Science. 2013.

http://www.garlandscience.com/product/isbn/9780815344544

TEXTBOOKS ABOUT BIOMATERIALS FOR FURTHER READING

Biomaterials Science: An Introduction to Materials in Medicine (Third Edition). Edited by Buddy Ratner, Allan S. Hoffman, Fredrick J. Schoen, and Jack E. Lemons. 3rd edition. Massachusetts: Elsevier. 2013.

Temenoff, J.S. and Mikos, A.G. Biomaterials: The Intersection of Biology and Materials Science. 1st edition. London: Pearson Science. 2008.

Williams, D.F. Essential Biomaterial Science (Cambridge Texts in Biomedical Engineering). 1st edition. Cambridge: Cambridge University Press 2014.

TEXTBOOKS ABOUT TISSUE ENGINEERING FOR FURTHER READING

Saltzman, M. Tissue Engineering: Engineering Principles for the Design of Replacement Organs and Tissues. Oxford University Press. 1st edition. 2004.

Principles of Tissue Engineering. Edited by Robert Lanza, Robert Langer, and Joseph Vacanti. Academic Press. Fourth edition. 2013.

COMPANIES THAT 3D PRINT TISSUES

Organovo located in San Diego, California: http://organovo.com/.

Aspect Biosystems located in Vancouver, British Columbia: http://aspectbiosystems.com

REFERENCES

[1] Dzyubenko E, Gottschling C, Faissner A. Neuron-glia interactions in neural plasticity: contributions of neural extracellular matrix and Perineuronal nets. Neural Plast 2016;2016:1–14.

[2] Chan B, Leong K. Scaffolding in tissue engineering: general approaches and tissue-specific considerations. Eur Spine J 2008;17(4):467–79.

[3] Mouw JK, Ou G, Weaver VM. Extracellular matrix assembly: a multiscale deconstruction. Nat Rev Mol Cell Biol 2014;15(12):771–85.

[4] Matthys OB, Hookway TA, McDevitt TC. Design principles for engineering of tissues from human pluripotent stem cells. Curr Stem Cell Rep 2016;2(1):43–51.
[5] Mansbridge J. Commercial considerations in tissue engineering. J Anat 2006;209(4):527–32.
[6] Sethi MK, Zaia J. Extracellular matrix proteomics in schizophrenia and Alzheimer's disease. Anal Bioanal Chem 2016;1–16.
[7] Fujii K, Tsuji M, Murota K. Isolation of peripheral nerve collagen. Neurochem Res 1986;11(10):1439–46.
[8] Chen P, Cescon M, Bonaldo P. The role of collagens in peripheral nerve myelination and function. Mol Neurobiol 2015;52(1):216–25.
[9] Banerjee J, Shi Y, Azevedo HS. In vitro blood–brain barrier models for drug research: state-of-the-art and new perspectives on reconstituting these models on artificial basement membrane platforms. Drug Discov Today 2016;21(9):1367–86.
[10] Malhotra V, Erlmann P. The pathway of collagen secretion. Annu Rev Cell Dev Biol 2015;31:109–24.
[11] Su K, Wang C. Recent advances in the use of gelatin in biomedical research. Biotechnol Lett 2015;37(11):2139–45.
[12] Halper J, Kjaer M. Basic components of connective tissues and extracellular matrix: elastin, fibrillin, fibulins, fibrinogen, fibronectin, laminin, tenascins and thrombospondins. Adv Exp Med Biol 2014;31–47.
[13] Yip JW, Yip YPL. Changes in fibronectin distribution in the developing peripheral nervous system of the chick. Dev Brain Res 1990;51(1):11–8.
[14] Li Y-C, Tsai L-K, Wang J-H, Young T-H. A neural stem/precursor cell monolayer for neural tissue engineering. Biomaterials 2014;35(4):1192–204.
[15] Yamada M, Sekiguchi K. Chapter six—Molecular basis of laminin–integrin interactions. Curr Top Membr 2015;76:197–229.
[16] Komura T, Kato K, Konagaya S, Nakaji-Hirabayashi T, Iwata H. Optimization of surface-immobilized extracellular matrices for the proliferation of neural progenitor cells derived from induced pluripotent stem cells. Biotechnol Bioeng 2015;112(11):2388–96.
[17] Miyazaki T, Kawase E. Efficient and scalable culture of single dissociated human pluripotent stem cells using recombinant E8 fragments of human laminin isoforms. Curr Protoc Stem Cell Biol 2014;32:1C.18.1–8.
[18] Plantman S. Proregenerative properties of ECM molecules. Biomed Res Int 2013;2013:1–11.
[19] Mercier F, Schnack J, Chaumet MSG. Fractones: home and conductors of the neural stem cell niche. In: Neurogenesis in the adult brain I. New York: Springer; 2011. p. 109–33.
[20] Theocharis AD, Skandalis SS, Gialeli C, Karamanos NK. Extracellular matrix structure. Adv Drug Deliv Rev 2016;97:4–27.
[21] Tassler P, Dellon A, Canoun C. Identification of elastic fibres in the peripheral nerve. J Hand Surg Eur Vol 1994;19(1):48–54.
[22] Muiznieks LD, Weiss AS, Keeley FW. Structural disorder and dynamics of elastin this paper is one of a selection of papers published in this special issue entitled "Canadian Society of Biochemistry, Molecular & Cellular Biology 52nd Annual meeting-protein folding: Principles and diseases" and has undergone the Journal's usual peer review process. Biochem Cell Biol 2010;88(2):239–50.
[23] Green EM, Mansfield JC, Bell JS, Winlove CP. The structure and micromechanics of elastic tissue. Interface Focus 2014;4(2):20130058.

[24] Faissner A, Roll L, Theocharidis U. Tenascin-C in the matrisome of neural stem and progenitor cells. Mol Cell Neurosci 2016;S1044-7431(16):30222–6.
[25] Nandadasa S, Foulcer S, Apte SS. The multiple, complex roles of versican and its proteolytic turnover by ADAMTS proteases during embryogenesis. Matrix Biol 2014;35:34–41.
[26] Maeda N. Proteoglycans and neuronal migration in the cerebral cortex during development and disease. Front Neurosci 2015;9:98.
[27] Neelamegham S, Mahal LK. Multi-level regulation of cellular glycosylation: from genes to transcript to enzyme to structure. Curr Opin Struct Biol 2016;40:145–52.
[28] Smith PD, Coulson-Thomas VJ, Foscarin S, Kwok JC, Fawcett JW. "GAG-ing with the neuron": the role of glycosaminoglycan patterning in the central nervous system. Exp Neurol 2015;274:100–14.
[29] Masu M. Proteoglycans and axon guidance: a new relationship between old partners. J Neurochem 2016;139(Suppl 2):58–75.
[30] Gasimli L, Linhardt RJ, Dordick JS. Proteoglycans in stem cells. Biotechnol Appl Biochem 2012;59(2):65–76.
[31] Wang P, Ding K. Proteoglycans and glycosaminoglycans in misfolded proteins formation in Alzheimer's disease. Protein Pept Lett 2014;21(10):1048–56.
[32] Fraser J, Laurent T, Laurent U. Hyaluronan: its nature, distribution, functions and turnover. J Intern Med 1997;242(1):27–33.
[33] Preston M, Sherman LS. Neural stem cell niches: critical roles for the Hyaluronan-based extracellular matrix in neural stem cell proliferation and differentiation. Front Biosci (Schol Ed) 2012;3:1165.
[34] Khaing ZZ, Milman BD, Vanscoy JE, Seidlits SK, Grill RJ, Schmidt CE. High molecular weight hyaluronic acid limits astrocyte activation and scar formation after spinal cord injury. J Neural Eng 2011;8(4):046033.
[35] Dzwonek J, Wilczynski GM. CD44: molecular interactions, signaling and functions in the nervous system. Front Cell Neurosci 2014;9:175.
[36] Baldwin KT, Giger RJ. Insights into the physiological role of CNS regeneration inhibitors. Front Mol Neurosci 2015;8:1–8.
[37] Mizumoto S, Yamada S, Sugahara K. Molecular interactions between chondroitin–dermatan sulfate and growth factors/receptors/matrix proteins. Curr Opin Struct Biol 2015;34:35–42.
[38] Dyck SM, Karimi-Abdolrezaee S. Chondroitin sulfate proteoglycans: key modulators in the developing and pathologic central nervous system. Exp Neurol 2015;269:169–87.
[39] Galtrey CM, Fawcett JW. The role of chondroitin sulfate proteoglycans in regeneration and plasticity in the central nervous system. Brain Res Rev 2007;54(1):1–18.
[40] Oduah EI, Linhardt RJ, Sharfstein ST. Heparin: past, present, and future. Pharmaceuticals 2016;9(3):38.
[41] Sakiyama-Elbert SE. Incorporation of heparin into biomaterials. Acta Biomater 2014;10(4):1581–7.
[42] Yamaguchi Y. Heparan sulfate proteoglycans in the nervous system: their diverse roles in neurogenesis, axon guidance, and synaptogenesis. Semin Cell Dev Biol 2001;12(2):99–106. Elsevier.
[43] Kraushaar DC, Dalton S, Wang L. Heparan sulfate: a key regulator of embryonic stem cell fate. Biol Chem 2013;394(6):741–51.
[44] Mercier F. Fractones: extracellular matrix niche controlling stem cell fate and growth factor activity in the brain in health and disease. Cell Mol Life Sci 2016;1–14.

[45] Funderburgh JL. Mini review keratan sulfate: structure, biosynthesis, and function. Glycobiology 2000;10(10):951–8.
[46] Miller B, Sheppard AM, Pearlman AL. Developmental expression of keratan sulfate-like immunoreactivity distinguishes thalamic nuclei and cortical domains. J Comp Neurol 1997;380(4):533–52.
[47] Zhang H, Uchimura K, Kadomatsu K. Brain keratan sulfate and glial scar formation. Ann N Y Acad Sci 2006;1086(1):81–90.
[48] Deepa SS, Carulli D, Galtrey C, Rhodes K, Fukuda J, Mikami T, et al. Composition of perineuronal net extracellular matrix in rat brain a different disaccharide composition for the net-associated proteoglycans. J Biol Chem 2006;281(26):17789–800.
[49] Medberry CJ, Crapo PM, Siu BF, Carruthers CA, Wolf MT, Nagarkar SP, et al. Hydrogels derived from central nervous system extracellular matrix. Biomaterials 2013;34(4):1033–40.
[50] Haggerty AE, Marlow MM, Oudega M. Extracellular matrix components as therapeutics for spinal cord injury. Neurosci Lett 2016;S0304-3940(16):30736–44.
[51] Kawano H, Kimura-Kuroda J, Komuta Y, Yoshioka N, Li HP, Kawamura K, et al. Role of the lesion scar in the response to damage and repair of the central nervous system. Cell Tissue Res 2012;349(1):169–80.
[52] Gonzalez-Perez F, Udina E, Navarro X. Extracellular matrix components in peripheral nerve regeneration. Int Rev Neurobiol 2013;108:257–75.
[53] Gause I, Sivak WN, Marra KG. The role of chondroitinase as an adjuvant to peripheral nerve repair. Cells Tissues Organs 2015;200(1):59–68.
[54] Gao X, Wang Y, Chen J, Peng J. The role of peripheral nerve ECM components in the tissue engineering nerve construction. Rev Neurosci 2013;24(4):443–53.
[55] Alovskaya A, Alekseeva T, Phillips J, King V, Brown R. Fibronectin, collagen, fibrin-components of extracellular matrix for nerve regeneration. Tissue Eng 2007;3:1–26.
[56] De Luca AC, Lacour SP, Raffoul W, Di Summa PG. Extracellular matrix components in peripheral nerve repair: how to affect neural cellular response and nerve regeneration? Neural Regen Res 2014;9(22):1943.
[57] Yamashita N, Kuruvilla R. Neurotrophin signaling endosomes: biogenesis, regulation, and functions. Curr Opin Neurobiol 2016;39:139–45.
[58] Levi-Montalcini R. The nerve growth factor: thirty-five years later. Biosci Rep 1987;7(9):681–99.
[59] Binder DK, Scharfman HE. Mini review. Growth Factors 2004;22(3):123–31.
[60] Hyman C, Juhasz M, Jackson C, Wright P, Ip NY, Lindsay RM. Overlapping and distinct actions of the neurotrophins BDNF, NT-3, and NT-4/5 on cultured dopaminergic and GABAergic neurons of the ventral mesencephalon. J Neurosci 1994;14(1):335–47.
[61] Barbacid M. The Trk family of neurotrophin receptors. J Neurobiol 1994;25(11):1386–403.
[62] Dechant G, Barde Y-A. The neurotrophin receptor p75NTR: novel functions and implications for diseases of the nervous system. Nat Neurosci 2002;5(11):1131–6.
[63] Morrison RS, Keating R, Moskal J. Basic fibroblast growth factor and epidermal growth factor exert differential trophic effects on CNS neurons. J Neurosci Res 1988;21(1):71–9.
[64] Funa K, Sasahara M. The roles of PDGF in development and during neurogenesis in the normal and diseased nervous system. J NeuroImmune Pharmacol 2014;9(2):168–81.
[65] Rimkus TK, Carpenter RL, Qasem S, Chan M, Lo H-W. Targeting the sonic hedgehog signaling pathway: review of smoothened and GLI inhibitors. Cancer 2016;8(2):22.

[66] Lai K, Kaspar BK, Gage FH, Schaffer DV. Sonic hedgehog regulates adult neural progenitor proliferation in vitro and in vivo. Nat Neurosci 2003;6(1):21–7.

[67] Fu M, Lui VCH, Sham MH, Pachnis V, Tam PKH. Sonic hedgehog regulates the proliferation, differentiation, and migration of enteric neural crest cells in gut. J Cell Biol 2004;166(5):673–84.

[68] Wall ST, Saha K, Ashton RS, Kam KR, Schaffer DV, Healy KE. Multivalency of sonic hedgehog conjugated to linear polymer chains modulates protein potency. Bioconjug Chem 2008;19(4):806–12.

[69] Bhattacharyya BJ, Banisadr G, Jung H, Ren D, Cronshaw DG, Zou Y, et al. The chemokine stromal cell-derived factor-1 regulates GABAergic inputs to neural progenitors in the postnatal dentate gyrus. J Neurosci 2008;28(26):6720–30.

[70] Davy A, Soriano P. Ephrin signaling in vivo: Look both ways. Dev Dyn 2005;232(1):1–10.

[71] Egea J, Klein R. Bidirectional Eph–ephrin signaling during axon guidance. Trends Cell Biol 2007;17(5):230–8.

[72] Conway A, Schaffer DV. Biomaterial microenvironments to support the generation of new neurons in the adult brain. Stem Cells 2014;32(5):1220–9.

[73] Maden M. Role and distribution of retinoic acid during CNS development. Int Rev Cytol 2001;209:1–77.

[74] Bain G, Kitchens D, Yao M, Huettner JE, Gottlieb DI. Embryonic stem cells express neuronal properties in vitro. Dev Biol 1995;168(2):342–57.

[75] Sinha S, Chen JK. Purmorphamine activates the hedgehog pathway by targeting smoothened. Nat Chem Biol 2006;2(1):29–30.

[76] Whitemarsh RC, Pier CL, Tepp WH, Pellett S, Johnson EA. Model for studying *Clostridium botulinum* neurotoxin using differentiated motor neuron-like NG108-15 cells. Biochem Biophys Res Commun 2012;427(2):426–30.

[77] Stanton BZ, Peng LF. Small-molecule modulators of the sonic hedgehog signaling pathway. Mol BioSyst 2010;6(1):44–54.

[78] Stacpoole SR, Bilican B, Webber DJ, Luzhynskaya A, He XL, Compston A, et al. Efficient derivation of NPCs, spinal motor neurons and midbrain dopaminergic neurons from hESCs at 3% oxygen. Nat Protoc 2011;6(8):1229–40.

[79] Gonzalez R, Garitaonandia I, Abramihina T, Wambua GK, Ostrowska A, Brock M, et al. Deriving dopaminergic neurons for clinical use. A practical approach. Sci Report 2013;3:1463.

[80] Robinson M, Yau S, Sun L, Gabers N, Bibault E, Christie BR, et al. Optimizing differentiation protocols for producing dopaminergic neurons from human induced pluripotent stem cells for tissue engineering applications. Biomark Insights 2015;1(Suppl):61–70.

[81] Keung AJ, Kumar S, Schaffer DV. Presentation counts: microenvironmental regulation of stem cells by biophysical and material cues. Annu Rev Cell Dev Biol 2010;26:533–56.

[82] Tian L, Prabhakaran MP, Ramakrishna S. Strategies for regeneration of components of nervous system: scaffolds, cells and biomolecules. Regener Biomat 2015;rbu017.

[83] Agbay A, Edgar JM, Robinson M, Styan T, Wilson K, Schroll J, et al. Biomaterial strategies for delivering stem cells as a treatment for spinal cord injury. Cells Tissues Organs 2016;202(1–2):42–51.

[84] Wang T-Y, Forsythe JS, Nisbet DR, Parish CL. Promoting engraftment of transplanted neural stem cells/progenitors using biofunctionalised electrospun scaffolds. Biomaterials 2012;33(36):9188–97.

[85] Xie J, MacEwan MR, Schwartz AG, Xia Y. Electrospun nanofibers for neural tissue engineering. Nanoscale 2010;2(1):35–44.
[86] Xie JW, Willerth SM, Li XR, Macewan MR, Rader A, Sakiyama-Elbert SE, et al. The differentiation of embryonic stem cells seeded on electrospun nanofibers into neural lineages. Biomaterials 2009;30(3):354–62.
[87] Mahairaki V, Lim SH, Christopherson GT, Xu L, Nasonkin I, Yu C, et al. Nanofiber matrices promote the neuronal differentiation of human embryonic stem cell-derived neural precursors in vitro. Tissue Eng A 2010;17(5–6):855–63.
[88] Spivey EC, Khaing ZZ, Shear JB, Schmidt CE. The fundamental role of subcellular topography in peripheral nerve repair therapies. Biomaterials 2012;33(17):4264–76.
[89] Yang Y, Leong KW. Nanoscale surfacing for regenerative medicine. Wiley Interdiscipl Rev Nanomed Nanobiotechnol 2010;2(5):478–95.
[90] Lee MR, Kwon KW, Jung H, Kim HN, Suh KY, Kim K, et al. Direct differentiation of human embryonic stem cells into selective neurons on nanoscale ridge/groove pattern arrays. Biomaterials 2010;31(15):4360–6.
[91] Mohtaram NK, Ko J, King C, Sun L, Muller N, Jun MB, et al. Electrospun biomaterial scaffolds with varied topographies for neuronal differentiation of human-induced pluripotent stem cells. J Biomed Mater Res A 2015;103(8):2591–601.
[92] Purcell EK, Naim Y, Yang A, Leach MK, Velkey JM, Duncan RK, et al. Combining topographical and genetic cues to promote neuronal fate specification in stem cells. Biomacromolecules 2012;13(11):3427–38.
[93] Jiang X, Cao HQ, Shi LY, Ng SY, Stanton LW, Chew SY. Nanofiber topography and sustained biochemical signaling enhance human mesenchymal stem cell neural commitment. Acta Biomater 2012;8(3):1290–302 [Research Support, Non-U.S. Gov't].
[94] Liu T, Xu JY, Chan BP, Chew SY. Sustained release of neurotrophin-3 and chondroitinase ABC from electrospun collagen nanofiber scaffold for spinal cord injury repair. J Biomed Mater Res A 2012;100A(1):236–42.
[95] Low WC, Rujitanaroj PO, Wang F, Wang J, Chew SY. Nanofiber-mediated release of retinoic acid and brain-derived neurotrophic factor for enhanced neuronal differentiation of neural progenitor cells. Drug Deliv Transl Res 2015;5(2):89–100.
[96] Mohtaram NK, Ko J, Agbay A, Rattray D, O'Neill P, Rajwani A, et al. Development of a glial cell-derived neurotrophic factor-releasing artificial dura for neural tissue engineering applications. J Mater Chem B 2015;3:7974–85.
[97] Horton ER, Humphries JD, James J, Jones MC, Askari JA, Humphries MJ. The integrin adhesome network at a glance. J Cell Sci 2016;jcs. 192054.
[98] Wang H, Luo X, Leighton J. Extracellular matrix and integrins in embryonic stem cell differentiation. Biochem Insights 2015;8(Suppl 2):15.
[99] Li Z, Lee H, Zhu C. Molecular mechanisms of mechanotransduction in integrin-mediated cell-matrix adhesion. Exp Cell Res 2016;349(1):85–94.
[100] Engler AJ, Sen S, Sweeney HL, Discher DE. Matrix elasticity directs stem cell lineage specification. Cell 2006;126(4):677–89.
[101] Saha K, Keung AJ, Irwin EF, Li Y, Little L, Schaffer DV, et al. Substrate modulus directs neural stem cell behavior. Biophys J 2008;95(9):4426–38.
[102] Chevalier N, Gazguez E, Bidault L, Guilbert T, Vias C, Vian E, et al. How tissue mechanical properties affect enteric neural crest cell migration. Sci Report 2016;349(1): 85–94.

[103] Matyash M, Despang F, Ikonomidou C, Gelinsky M. Swelling and mechanical properties of alginate hydrogels with respect to promotion of neural growth. Tissue Eng Part C Methods 2013;20(5):401–11.

[104] Zhou W, Stukel JM, Cebull HL, Willits RK. Tuning the mechanical properties of poly (ethylene glycol) microgel-based scaffolds to increase 3D Schwann cell proliferation. Macromol Biosci 2015;16(4):535–44.

[105] Downing TL, Soto J, Morez C, Houssin T, Fritz A, Yuan F, et al. Biophysical regulation of epigenetic state and cell reprogramming. Nat Mater 2013;12(12):1154–62.

[106] Budday S, Nay R, de Rooij R, Steinmann P, Wyrobek T, Ovaert TC, et al. Mechanical properties of gray and white matter brain tissue by indentation. J Mech Behav Biomed Mater 2015;46:318–30.

[107] Franze K. The mechanical control of nervous system development. Development 2013;140(15):3069–77.

[108] Mekhail M, Tabrizian M. Injectable chitosan-based scaffolds in regenerative medicine and their clinical translatability. Adv Healthc Mater 2014;3(10):1529–45.

CHAPTER

Natural biomaterials for engineering neural tissue from stem cells

5

1 INTRODUCTION

Biomaterials are defined as materials that interact with the body to perform or supplement its natural functions [1]. These materials must also demonstrate the property of biocompatibility, meaning that they do not trigger a systemic immune response from the body after implantation. This chapter focuses on the different types of biomaterials obtained from natural sources. The two major classes of biomaterials include proteins and polysaccharides, which are also biopolymers. These molecules play an important role in ensuring proper cell function as they compose different parts of the extracellular matrix. Their intrinsic properties makes them an excellent starting point for engineering tissues from stem cells as they mimic the properties of the extracellular matrix, which was described in depth in Chapter 4 [2]. These materials are often bioactive and their properties are similar to those possessed by soft tissues found in vivo. However, their mechanical properties are often limited and it can be difficult to tune these mechanical properties, unlike synthetic biomaterials discussed in Chapter 6.

Sourcing large quantities of these proteins and polysaccharides can be challenging, especially when trying to scale up production. They can be obtained from animals if proper processing is performed to ensure these materials will not trigger an immune response in humans [3]. A recent review details the major issues that need to be addressed before such animal-derived materials can be translated for clinical applications [4]. These issues include cytotoxicity, degradation rate, pathogenicity, and potential to trigger a significant immune response from the host. More on specific testing methods for assessing these parameters can be found by reading the textbooks mentioned in the Resources section. An additional limitation of natural biomaterials is that their material characteristics can vary depending on how it is sourced, leading to variability in their biological properties between batches. Overall, these limitations are acceptable as these scaffolds provide excellent environments for supporting stem cell growth and differentiation into neural tissue as detailed in this chapter. Combining natural and synthetic biomaterials can also address the limitations associated with natural biomaterials as detailed in Chapter 6.

In this chapter, Section 2 explores examples of using scaffolds made predominately from one type of biomolecule while Section 3 looks at examples of using decellularized tissues as scaffolds for promoting neural regeneration and stem cell

Engineering Neural Tissue from Stem Cells. http://dx.doi.org/10.1016/B978-0-12-811385-1.00005-4

differentiation. Section 4 offers suggestions for future work in how to improve the properties of scaffolds fabricated from natural biomaterials.

2 PURE BIOMATERIALS

This section will focus on biomaterial scaffolds composed from mainly of one type of biological molecule. The properties of each molecule are reviewed, along with the associated advantages and disadvantages, followed by relevant examples to neural tissue engineering with an emphasis on applications using stem cells.

2.1 PROTEINS

Here, recent advances in using protein-based biomaterial scaffolds for repairing the damaged nervous system are reviewed by discussing selected relevant examples. The most commonly used protein scaffolds include collagen, fibrin, fibronectin, and silk. These proteins can be mixed together to further enhance the bioactive properties of naturally derived scaffolds and certain examples of these combinatorial scaffold materials will be reviewed in this chapter as well.

2.1.1 Collagen

As detailed in Chapter 4, the extracellular matrix contains significant quantities of collagen, making it a logical starting point for engineering neural tissue. While brain tissue does not contain much collagen, other tissues such as skin, ligament, and tendons, contain significant amounts of collagen. These tissues can be processed to obtain collagen for tissue engineering applications. Collagen can be sourced from both human and nonhuman sources and it is typically sourced from pigs, horses, and cows. Collagen scaffolds are already used clinically to promote wound healing, usually in the context of injuries to the skin [5]. Also, collagen can be injected cosmetically to eliminate wrinkles and to plump lips [6]. Thus, the major biocompatibility testing required for clinical use has already been conducted for this protein, enhancing its desirability as a biomaterial scaffold for tissue engineering applications. A recent review detailed how different properties of collagen scaffolds can be further optimized to improve its effectiveness as a scaffold material [7]. In an example of this process, the surface properties of collagen can be controlled so that these scaffolds promote regeneration of peripheral nerves instead of leading to scarring [8]. Collagen comes in many different forms with type I collagen being the most commonly used version for tissue engineering as discussed in the Chapter 4.

Early work in 2000 demonstrated that collagen could support the culture and differentiation of primary neural precursor cells isolated directly from the brain of 13-day-old rats when cultured in neural basal media containing basic fibroblast growth factor (bFGF) [9]. These cells differentiated into Tuj1 positive neurons and responded to treatment with glutamate by releasing calcium ions, suggesting they were functional neurons. Similar work also confirmed the ability of these cells to form functional circuits inside of three-dimensional (3D) collagen constructs (Fig. 1)

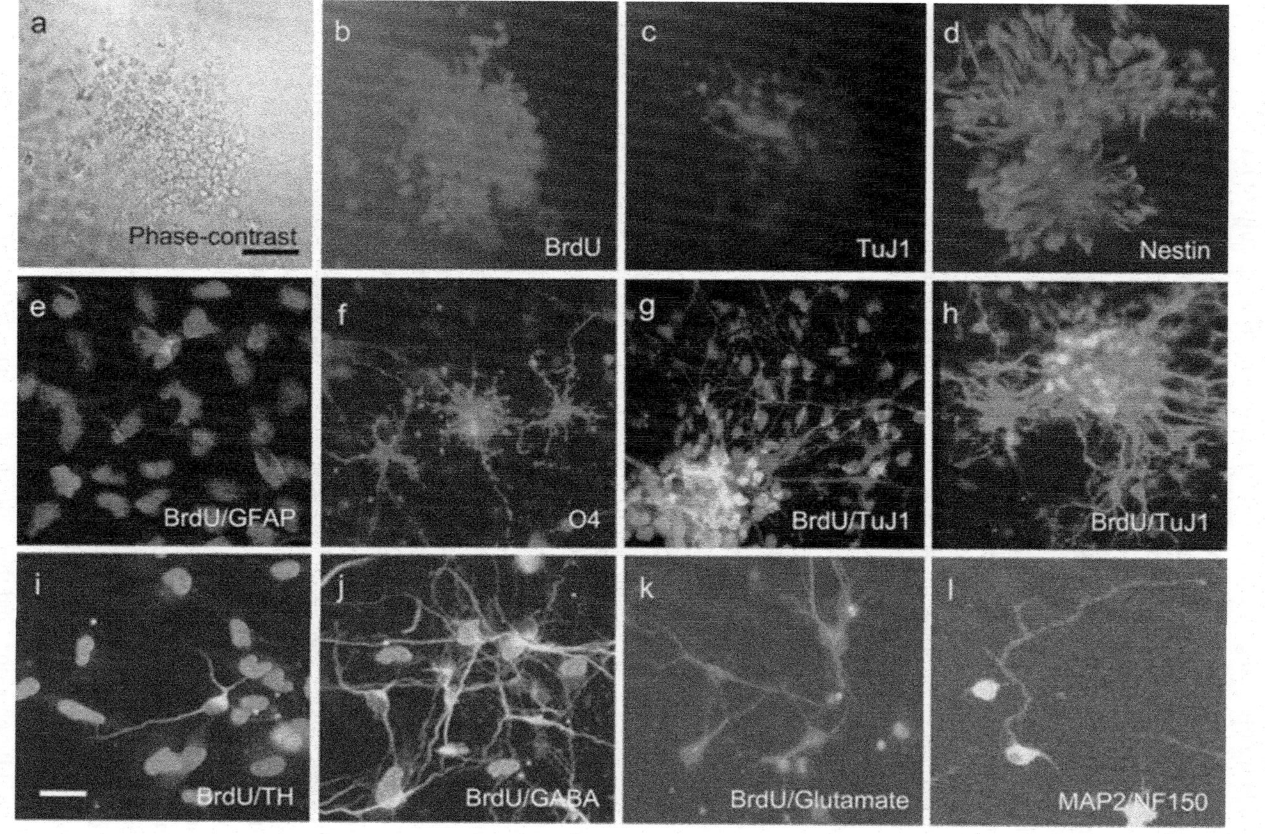

FIG. 1

Collagen-entrapped cortical or subcortical progenitor cells differentiate into neurons, astrocytes and oligodendrocytes. Progenitor cells were expanded in a collagen gel for 3, 5, 7, 14, and 21 days, then immunostained for neuronal (TuJ1) and glial (GFAP, O4) markers, BrdU, nestin, neurofilaments (MAP2/NF150), and neurotransmitters or enzymes (GABA, glutamate, tyrosine hydroxylase, TH). At day 5, many of the cells visualized in phase contrast (A) are actively proliferating (BrdU+, B) and nestin+(D), but fewer are neuronal (TuJ1+, C). At day 14 in culture, a variety of differentiated cells appear. At higher magnification (E–L), individual progeny can be identified that are GFAP+ (E). Many cells are BrdU+. Some cells are oligodendrocytes (O4+, F). Cell migration occurs from a core of a proliferating cluster (BrdU+, G, H). Some of neurons are TH+ (I) and BrdU+ while others are either GABA+ (J) and BrdU+ or glutamate+(K) and BrdU+. Both dendrites (MAP2+) and axon (NF150+) are differentiated by neurons (L). Scale bar in $a = 150\,\mu m$ and applies to B–D; in $l = 50\,\mu m$ and applies to (E)–(L).

Reprinted with permission from Ma W, Fitzgerald W, Liu Q-Y, O'shaughnessy T, Maric D, Lin H, et al. CNS stem and progenitor cell differentiation into functional neuronal circuits in three-dimensional collagen gels. Exp Neurol 2004;190(2):276–88.

[10] as well as demonstrating how bFGF promotes the differentiation of rat neural stem cells into mature neurons when seeded inside 3D collagen gels [11]. Later work demonstrated how collagen gels could be used as microcarriers to culture rat neural stem cells in stirred bioreactors [12]. A 2007 study characterized the phenotypes generated from rat neural stem cells cultured inside of 3D collagen scaffolds and found that ~40% became neurons with ~50% becoming astrocytes and the rest becoming oligodendrocytes [13].

More recent studies have further confirmed that different formulations of 3D collagen scaffolds support the culture of rat neural stem cells [14,15]. This work found that the gels were supportive of neural differentiation while the sponges tended to promote aggregation of the stem cells. In 2010, it was shown that 3D collagen supported the culture of astrocytes derived from human neural stem cells, demonstrating the versatility of these scaffolds for a range of mammalian neural stem cells [16]. These astrocytes also promoted the regeneration of axons from dorsal root ganglia, suggesting they maintain their function in such 3D–engineered tissue constructs. In 2013, 3D collagens scaffolds were used to culture successfully neural-like cells derived from umbilical cord blood stem cells that appeared to form neural networks [17]. Further work has demonstrated that the stiffness of a collagen gel influences the migration rate of stem cells with stiffer substrates promoting migration of neuronal and glial cells [18]. The Liu group developed mathematical models to study how neural stem cells proliferate and function when seeded inside of 3D collagen scaffolds [19]. Such models are a valuable tool for performing simulations of cell behavior without the expense of performing experiments.

Collagen scaffolds also support the culture and differentiation of neural cells derived from pluripotent stem cells, in addition to neural stem cells. In fact, collagen is one of the major components found in Matrigel, the complex mixture of extracellular matrix components produced by mouse cancer cells. Matrigel was previously considered the gold standard matrix for culturing undifferentiated human pluripotent stem cells [20] before more defined matrices were identified. Thus, it is logical to explore how pluripotent stem cells seeded inside of collagen scaffolds can be manipulated to differentiate into functional neural tissue.

In 2006, Chen and colleagues evaluated two different collagen scaffold formulations (a porous sponge and a dense gel) to determine their effect on the differentiation of embryonic stem cells derived from Rhesus monkeys [21]. They found the porous scaffolds supported aggregate formation while the cells seeded inside of the gels formed gland-like structures. Collagen scaffolds also supported the culture and differentiation of mouse embryonic stem cells into neural lineages [22]. In other work on collagen scaffolds for pluripotent stem cell culture, altering the pore structure of collagen scaffolds as well as functionalizing them with laminin improved the growth of human-induced pluripotent stem cell (hiPSC)-derived neural progenitors cells seeded inside [23]. As discussed in Chapter 4, the stiffness of the collagen matrix plays an important role in differentiation, and these effects were characterized of hiPSC-derived embryoid bodies on collagen matrices where softer substrates encouraged neural differentiation [24].

Several studies have incorporated controlled drug delivery systems in such scaffolds to enable them to deliver bioactive cues for influencing stem cell behavior. A version of epidermal growth factor (EGF) functionalized to bind collagen was added to such scaffolds as a way to generate controlled release [25]. The controlled delivery of EGF enhanced cell proliferation inside of these 3D scaffolds. Another group used this approach to deliver bFGF to neural stem cell in vitro to enhance their proliferation [26]. The use of functionalized proteins designed to have an affinity interaction with such scaffolds will be discussed in depth in Chapter 7. In similar work, collagen scaffolds designed to release brain-derived neurotrophic factor (BDNF) and neurotrophin-3 (NT-3) supported survival and proliferation of neural stem cells compared to the individual factors alone and negative control collagen scaffolds [27]. Functionalized collagen scaffolds with laminin-derived peptides promoted an increase in neural stem cell survival inside of these gels, providing a different avenue for manipulating cell behavior [28]. Controlled delivery of BDNF from collagen scaffolds promotes the proliferation and differentiation of adult rat neural stem cells [29,30]. More recently, the effect of nerve growth factor (NGF) on promoting migration of neural stem cells seeded inside of collagen scaffolds was confirmed using microfluidics [31].

Several studies have demonstrated that collagen scaffolds effectively deliver different types of stem cells into the damaged nervous system in vivo by enhancing survival and differentiation. For example, collagen scaffolds could successfully deliver ammonitic stem cells into a rat traumatic brain injury model, where the transplanted cells enhanced neural regeneration [32]. A similar study used a collagen scaffold to deliver human mesenchymal stem cells into a traumatic brain injury model where this engineered tissue enhanced axonal density post injury [33]. Collagen scaffolds seeded with adult neural progenitors delivered into a traumatic brain injury model enabled the cells to survive and differentiate in vivo as well as migrate into the surrounding tissue [34]. The collagen scaffold enabled the cells to survive and mediate wound healing. Another study transplanted a nerve guidance conduit filled with collagen seeded with neural crest stem cells into a sciatic nerve injury model where this engineered tissue enhanced regeneration post injury [35]. More recent work indicated that an engineered tissue consisting of adipose-derived stem cells and a collagen scaffold could bridge critical sized defects (>5 mm) in a rat sciatic nerve injury model [36]. Also, collagen scaffolds delivering neural stem and bone marrow stromal cells improved outcomes in preclinical models of stroke [37,38]. Collagen alone and in combination with stem cells promoted better alignment of scar tissues in a preclinical spinal cord injury model [39]. Thus, it was not surprising when a 2013 study showed that using collagen functionalized with the EGF receptor to deliver rat neural stem cells into the damaged spinal cord resulted in enhanced axonal regeneration and functional recovery [40]. Another study that delivered neural stem cells using a collagen scaffold fabricated to contain heterogeneous pores into the injured spinal cord of rats where they also promoted functional recovery [41]. Other work on spinal cord injury used collagen microfibers to deliver neural stem cells into the injured spinal cord where they could survive and differentiate, but did not promote functional recovery [42]. In a different type of preclinical setting, hiPSC-derived neural progenitors were delivered successfully in the cochleae of guinea pigs where they differentiated into neurons [43].

Other groups have electrospun collagen into nanofibers, which support neural stem cell culture and their ability extend to neurites [44,45]. However, solution electrospinning methods often denature the protein structure, causing a lack of bioactivity. An interesting approach fabricated collagen microchannels that promoted the formation of 3D networks of astrocytes [46]. These tubes mimicked the tracts that neural progenitors migrate along in the brain and thus they could potentially serve as a mechanism for stimulating endogenous neural stem cell migration in vivo. Such nanofiber collagen scaffolds can also serve as substrates for the culture and differentiation of neural stem cells derived from the spinal cord [47]. These nanofiber scaffolds can also be pattern with protein gradients to direct stem behavior. For example, gradients of stromal-derived factor 1 (SDF1) were immobilized on collagen nanofibers using a collagen-binding domain. These gradient-patterned scaffolds induced migration of neural stem cells toward the regions with higher SDF1 concentrations, providing an intriguing strategy for engineering neural tissue [48]. Another intriguing form of collagen scaffolds encapsulates cells inside of microspheres as unique method for cell delivery. A recent study in 2013 demonstrated that such microsphere encapsulated oligodendrocyte progenitors had increased survival and differentiation into cells that could myelinate neurons [49]. This strategy could be applied in vivo to treat different types of neurological disorders where demyelination occurs.

Overall, these studies indicate how the natural properties of collagen make it an excellent scaffold material for cell delivery to the damaged nervous systems. Certain studies also demonstrated how combining growth factor delivery with these scaffolds can further enhance their bioactive properties and that these properties can control the behavior of stem cells seeded inside. One of the major limitations of collagen scaffolds is its relatively quick degradation rate, which impacts its ability to deliver physical and chemical cues over extended times. These properties can be modified through processing techniques and the use of cross-linking reagents as mentioned earlier in this section. Later in this section, combinations of collagen with other proteins will be discussed and in Chapter 6, studies that combine natural and synthetic biomaterials to generate multifunctional scaffolds for neural tissue engineering applications will be discussed.

2.1.2 Fibrin

Fibrin plays a critical role in the process of coagulation where it polymerizes to form blood clots [50]. A large biomolecule called fibrinogen is cleaved by the enzyme thrombin to generate fibrin monomers which then assemble into a scaffold that becomes stabilized by covalent cross-links generated by the enzyme Factor XIIIa. It is possible to isolate fibrinogen from a patient's own blood although it can also be sourced from animals in a manner similar to collagen. Thus, it is an attractive material for tissue engineering applications [51]. It can also be modified to contain drug delivery systems [52] and its mechanical properties can be varied through different chemical processes [53–57].

Pioneering work developing fibrin scaffolds and their novel derivatives include a study by Dr. Shelly Sakiyama-Elbert and Dr. Jeff Hubbell, who demonstrated such scaffolds could be functionalized with a novel heparin-binding delivery system. These functionalized scaffolds supported neural cell culture and promoted neurite extension of primary neural cultures isolated from chicken embryos [58–60]. When I was a graduate student at Washington University working under the guidance of Dr. Sakiyama-Elbert, one of my first projects was determining an optimal fibrin formulation for the culture of neural progenitors derived from mouse embryonic stem cells [55]. My work also showed that neural progenitors derived from mouse embryonic stem cells survived better inside of 3D scaffolds when cultured as intact embryoid bodies as opposed to dissociated cells. I was then able to use this fibrin scaffold as a platform for screening different growth factors, determining how different concentrations and combinations of these factors influence the differentiation of these neural progenitors into different percentages of neurons, oligodendrocytes, and astrocytes [61]. Finally, we translated this work on growth-mediated differentiation into engineered tissue consisting of mouse embryonic stem cell–derived embryoid body containing neural progenitors, an affinity-based drug delivery system and different growth factor combinations [62].

Further work from the Sakiyama-Elbert group evaluated the effects of this engineered tissue in a dorsal hemisection model of spinal cord injury in rats, where this tissue promoted the survival and differentiation of the cells into neurons and oligodendrocytes [63]. A follow-up study indicated that this engineered tissue could also promote functional recovery 4 weeks post injury [64]. Further work from her group used these scaffolds to transplant motor neuron progenitors derived from mouse embryonic stem cells [65]. These cells survived in the injured spinal cord and became neurons, oligodendrocytes, and astrocytes (Fig. 2). Finally, they combined the delivery of drugs to reduce the inhibitory nature of the injury site with such an engineered tissue, although the results were suboptimal as this combinatorial therapy resulted in reduced cell survival [66].

In addition to the Sakiyama-Elbert group, several other groups have used fibrin scaffolds to deliver cells to the injured spinal cord. In 2012, the Tuszynski group delivered human neural stem cells seeded inside of a fibrin scaffold containing several growth factors into a severe spinal cord injury [67,68]. These cells integrated into the injured spinal cord where they could grow over long distances and promote functional recovery. However, a follow-up replication study indicated that the cells survived and differentiated, but no functional recovery was observed [69]. A similar approach was used to deliver neural stem cells derived from human-induced pluripotent stem cells where the results showed good cell survival and integration, but not functional recovery [70].

Since starting my independent research group, we have explored how such fibrin scaffold could support the culture of induced pluripotent stem cells for neural tissue engineering. In 2015, we showed that neural progenitors derived from mouse-induced pluripotent stem cells could both survive and differentiate into neurons inside of 3D fibrin scaffolds in vitro [71]. This work demonstrated that a combination

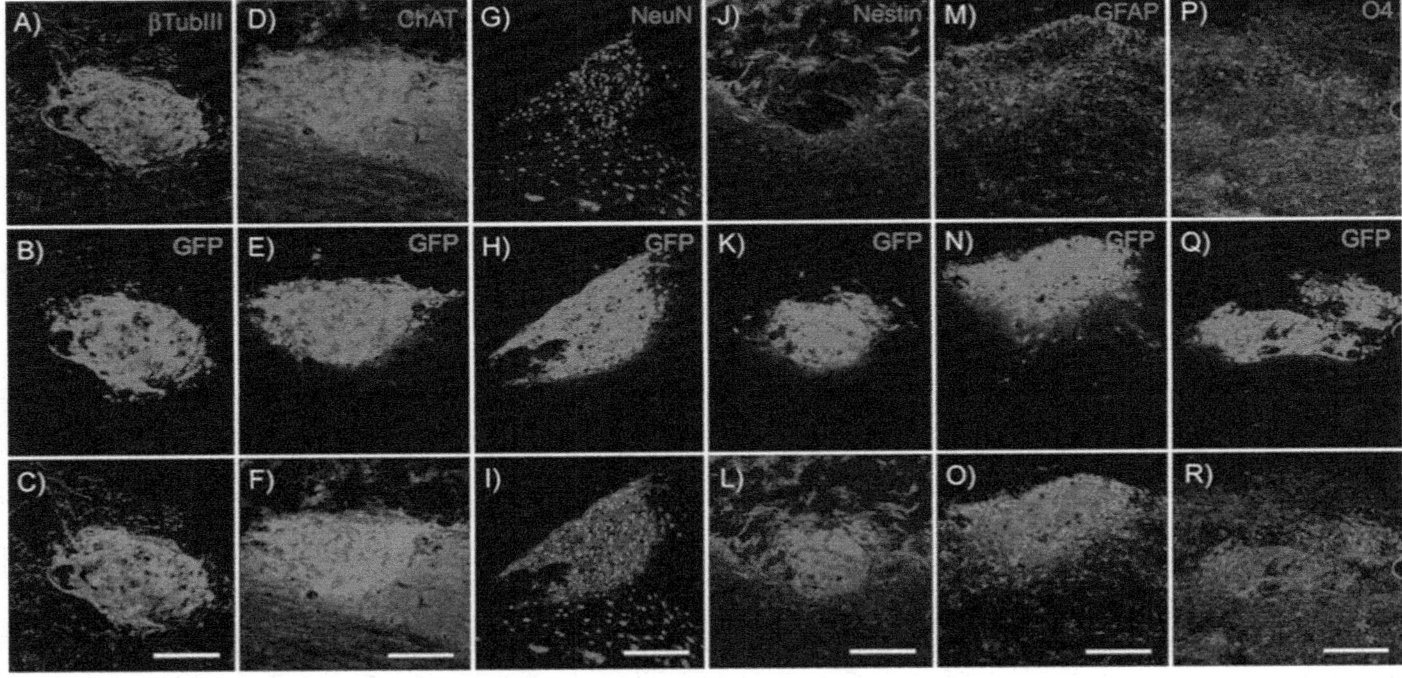

FIG. 2

Immunofluorescence images of cell differentiation within the injured spinal cord two weeks posttransplantation. (A) β-tubulin III labeling and corresponding (B) GFP expression in transplanted cells (Fibrin). (C) Merged view β-tubulin III and green fluorescent protein (GFP). (D) Choline acetyltransferase (ChAT) labeling and (E) GFP expression (DS+NT3+GDNF). (F) Merged view of ChAT and GFP. (G) NeuN labeling and (H) GFP expression (DS). (I) Merged view of NeuN and GFP. (J) Nestin labeling and (K) GFP expression (DS+NT3+GDNF). (L) Merged view of GFP and nestin. (M) GFAP labeling and (N) GFP expression (DS+NT3+PDGF). (O) Merged view of GFAP and GFP. (Q) O4 labeling and (R) GFP expression (DS+NT3+PDGF). (S) Merged view of O4 and GFP. Abbreviations: *βtubIII*, β-tubulin class III; *GFAP*, glial fibrillary acidic protein; *O4*, oligodendrocyte marker 4; *NeuN*, neuronal nuclei; *ChAT*, choline acetyltransferase; *GFP*, green fluorescent protein. Scale bar = 250 μM.

Reprinted with permission from McCreedy DA, Wilems TS, Xu H, Butts JC, Brown CR, Smith AW, et al. Survival, differentiation, and migration of high-purity mouse embryonic stem cell-derived progenitor motor neurons in fibrin scaffolds after sub-acute spinal cord injury. Biomater Sci 2014;2(11):1672–82.

of retinoic acid and purmorphamine generated more neurons from neural progenitors cultured in such 3D systems compared a protocol that only used retinoic acid as an agent for neuronal patterning. Encouraged by this study, we then examined the culture of neural progenitors derived from hiPSCs inside of such scaffolds [72]. These cells did not differentiate into neurons as effectively as the neural progenitors derived from mouse-induced pluripotent stem cells, and they also degraded the scaffolds more rapidly. The degradation is possibly due to the human cells being more effective at breaking down human fibrinogen compared to cells derived from mice. Ongoing work in my laboratory focuses on enhancing the differentiation of neural progenitors derived from induced pluripotent stem cells and increasing the stability of fibrin using cross-linking reagents such as genipin and protease inhibitors like aprotinin. Our recent publication also elucidated the differences in gene expression between hiPSC-derived neural progenitors cultured in 3D fibrin scaffolds compared to those cultured on two-dimensional (2D) laminin surfaces [73].

Other work has shown that fibrin scaffolds can direct mesenchymal stem cells to form neuron-like cells, demonstrating their versatile properties [74]. They can also support culture of endometrial stromal cell-derived oligodendrocyte progenitor cells, another interesting cell source [75]. Interestingly, human peripheral blood mononuclear cells cultured inside of fibrin scaffolds express neuronal markers as well [76]. Such an engineered tissue could be derived entirely from a patient's blood as a source for both the cells and fibrin. In a different study, fibrin scaffolds in combination with bFGF treatment were used to deliver bone marrow-derived stem cells into an animal model of traumatic brain injury where this unique combination promoted a reduction in infarct volume and apoptosis—leading to better functional outcomes [77]. Injecting fibrin glue containing induced pluripotent stem cells into the infarction caused by stroke resulted in improved functional recovery as well [78].

Recently, two unique approaches developed fibrin scaffolds for engineering brain tissue. In the first study, the Demicri group used bioacoustic forces to position neural progenitors derived from human embryonic stem cells into precise locations inside of 3D fibrin constructs, replicating the tissue found in the brain [79]. The strategy is interesting because it enables the placement of different cells into precise positions within a construct, which can be used to engineer different cellular compositions for each layer of tissue [80]. The second approach used fibrin scaffolds as a way to engineer a pathway for the endogenous neural stem cells of the brain to migrate along [81]. This fibrin pathway, which was loaded with growth factors, could direct endogenous stem cells into other regions of the brain, providing an interesting alternative way of engineering neural repair by harnessing the resident stem cells of the brain. Overall, the studies detailed in this section illustrate the effectiveness of fibrin as a scaffold for supporting stem cell differentiation into neural tissue.

2.1.3 Fibronectin

Fibronectin plays a critical role in the extracellular matrix and in repair of the nervous system as detailed in Chapter 4 [82]. This protein possesses an extremely high molecular weight as it consists of a dimer with each subunit weighing ~440 kDa.

It contains binding sites for integrins that enable the cells present in the nervous system to adhere. Thus, it is often used in combination with other scaffold materials as a way to enhance their ability to bind cells [83]. Fibronectin also promotes migration of stem cells, making it attractive for cell transplantation studies [84]. The studies using fibronectin in combination with other materials will be discussed in Section 2.3, while this section will focus on scaffolds composed of primarily fibronectin. 2D layers of fibronectin bind neural stem cells through the activity of integrins. Cells can proliferate on these surfaces, providing an alternative to culturing these stem cells in non-adherent conditions [85]. Fibronectin surfaces can also serve as substrates for culturing neural stem cells derived from mammalian pluripotent stem cells. It also increases neural stem cell survival when added to cell culture media, suggesting its potential as cell culture supplement [86].

A study comparing the properties of various components of the extracellular matrix determined that fibronectin induced the most positive effects, including promoting cell survival and growth, on oligodendrocyte progenitor cells, providing more evidence of its bioactive properties [87]. Human peripheral blood mononuclear cells cultured in fibronectin exhibit neuronal properties with similar observations being made for cord blood-derived neural cells, indicating its versatility [88,89]. Finally, a recent study found that mesenchymal stem cells secrete significant amount of fibronectin, which in turn enhances axonal extension, promoting regeneration in a preclinical spinal cord injury model [90]. While these studies did not use fibronectin as a 3D scaffold, they do illustrate its importance as a bioactive cue for engineering neural tissue. Fibronectin is commonly used in conjunction with other materials for neural tissue engineering applications, and these studies will be discussed later in this chapter and Chapter 6.

2.1.4 Silk

Although it is more commonly associated as material used to manufacture clothing, biomaterials scientists have evaluated the fibrous protein silk for a wide range of medical applications [91]. Both silkworms and spiders produce silk fibers. These fibers can then be processed into 3D scaffolds, which take a variety of forms, including sponges, gels, and films. Silk demonstrates excellent biocompatibility and it exhibits a large range of mechanical properties, which vary on the method of processing. It also can provide controlled release of different factors, which is an attractive property when working with stem cells [92]. This section will discuss how silk scaffolds can be combined with different types of stem cells to engineer neural tissue.

Silk nerve guidance conduits containing aligned fibers seeded with rat mesenchymal stem cells promoted regeneration and functional recovery after 12 weeks in a rat sciatic nerve injury model [93], demonstrating its biocompatibility with the nervous system and its bioactive properties. In a related study, the effect of changing the topography of silk scaffolds on the differentiation of human embryonic stem cell-derived neural precursors was determined [94]. They found that the optimal diameter of these fibers was 400 nm and that aligning electrospun fibers with this diameter enhanced neuronal differentiation and neurite extension. These results were supported by another similar study looking at the effect of features in same size range [95]. Another study used electrospun silk fibers for successfully culturing Schwann cells as they could grow and proliferate on these scaffolds [96]. Later work showed the recombinant spider silk matrices could support the adhesion of rat neural stem cells, which differentiated into neurons and astrocytes through supplementing the media with growth factors [97]. Such electrospun silk scaffolds can also direct mesenchymal stem cells to form neuron-like cells [98]. In a different approach to scaffold fabrication, hydrogels with tunable stiffness were fabricated from silk nanofibers that supported the culture of neural stem cells [99]. These hydrogels can also be functionalized with peptides to enhance their properties further. For example, silk hydrogels containing the IKVAV peptide sequence taken from laminin increased cell viability and neuronal differentiation of human neural stem cells compared to control hydrogels consisting only of silk [100]. Other work has also confirmed the suitability of silk scaffolds for culturing and differentiating human neural progenitors [101]. This set of studies illustrates how silk serves an attractive alternative scaffold material compared to the more commonly used proteins, collagen, and fibrin.

2.1.5 Laminin

Laminin plays a critical role in promoting axonal extension in the developing nervous system as detailed in Chapter 4. The peptide sequence IKVAV is derived from one of domains found in laminin. This peptide domain is commonly used to functionalize biomaterials to enable cell adhesion [102]. A comparison of four different 2D surface coatings, including laminin, Matrigel, poly-L-ornithine, and fibronectin, found that laminin was the most effective at promoting the expansion and differentiation of both mouse and human neural progenitors into mature neural phenotypes [103]. Laminin also promoted the greatest amount of neurite extension from the neurons generated in these seeded cultures. These properties indicate its utility for engineering neural tissue from stem cells. Further work has used 2D patterning of laminin to direct the alignment of different types of neural cells, including astrocytes and adult rat neural progenitors [104]. The coculture of these progenitors with astrocytes also induced a significant increase in the number of cells that differentiated into neurons. Laminin surfaces can also be used in combination with growth factor treatment to regulate neural progenitor behavior [105]. Most scaffolds use the properties of laminin in combination with other biomaterials or use its active domains as detailed in this chapter or as a part of synthetic matrices as detailed in Chapter 6.

2.2 POLYSACCHARIDES

Polysaccharides consist of sugar monomers chemically linked together by covalent bonds to produce bioactive chains. Like proteins, polysaccharides are a diverse set of molecules with bioactive properties and perform complex roles in the extracellular matrix, where they influence cellular behavior. Many of these polysaccharides play an important role in developing and mature nervous system [106]. However, some of the polysaccharides discussed in this section are sourced from other species as detailed below. This section discusses some of the most commonly used polysaccharides for applications in neural tissue engineering, including agarose, alginate, chitosan, methylcellulose, nitrocellulose, dextran, and hyaluronan.

2.2.1 Agarose

Seaweed contains the polysaccharide agarose, which also makes up agar [107]. It is commonly used to produce the gels used during the electrophoresis process, which separates molecules by size. Thus, it readily polymerizes into a hydrogel whose properties can be tuned by varying different parameters like the concentration of agarose used. Dr. Ravi Bellamkonda (now the Vinik Dean of the Pratt School of Engineering at Duke University) and his colleagues working in the group of Dr. Patrick Aebischer demonstrated that 3D agarose hydrogels supported neurite outgrowth from chick dorsal root ganglia in a concentration dependent manner, suggesting that agarose was a promising material for use when engineering neural tissue [108]. However, agarose does not contain natural binding sites for mammalian cells to adhere as it is derived from seaweed. In fact, one study demonstrated that coating tissue culture plastic with agarose prevented neural stem cells from binding and they used this coating as a way to prevent stem cell attachment and induce formation of neurospheres [109]. Other work has used agarose-coated microwells for generating embryoid bodies containing neural progenitors from human embryonic stem cells due to their resistance to cell adhesion [110]. Interestingly, one study showed that agarose increased the differentiation of rat neural stem cells into neurons despite its nonadhesive properties [111]. Due to these properties, agarose is often modified with bioactive proteins or peptide domains to enhance its ability to bind stem cells. One of the advantages of using agarose is that it can be easily manipulated into different shapes and thus can be structured to have channels for directing axonal outgrowth from neuronal cells [112]. Work from Dr. Molly Shoichet's group at the University of Toronto shows that modifying agarose with peptides containing glycine-arginine-glycine-aspartic acid-serine (GRGDS) resulted in a hydrogel that supported the culture of neural stem cells [113]. These scaffolds were then modified to deliver platelet-derived growth factor AA (PDGF-AA), which promoted the differentiation of these stem cells into oligodendrocytes. Further work from the Shoichet lab conducted by Dr. Ryan Wylie showed that EGF could be patterned into agarose scaffolds in three dimensions, allowing for the precise localization of cues within these scaffolds [114]. The Wu group incorporated agarose hydrogels into their microfluidic devices as a method for assessing the migration of neural stem cells using this system [115]. They identified overexpression of the EGF receptor as playing a critical role in this migration process. Overall, agarose has some utility for engineering neural tissue from stem cells, but it often is used in combination with other components to improve its bioactivity.

2.2.2 Alginate

The cell walls of brown algae contain the polysaccharide alginate, which forms a viscous solution when mixed with water [116]. Some bacteria also secrete alginate as one of the components of their biofilms [117]. After being isolated from these sources, alginate can be polymerized into hydrogels using calcium cross-linking or through the use of chemistries that form covalent bonds. Alginate hydrogels containing aligned channels promoted axonal regeneration both in vitro and in a rat spinal cord injury model, suggesting their compatibility with neural cells [118]. These scaffolds tend last for extended time periods in vivo compared to other naturally derived biomaterials as the human body does not secrete the enzymes necessary to degrade alginate.

Alginate beads can encapsulate neural stem cells for scale-up culture using bioreactors as the process of cross-linking is reversible, allowing the encapsulated cells to be recovered post culture [119]. Work from Dr. Randolph Ashton (now at the University of Wisconsin-Madison) and colleagues conducted under the supervision of Dr. David Schaffer and Dr. Ravi Kane showed that the degradation rate of alginate scaffolds could be tuned by incorporation of poly(lactic-*co*-glycolic) microspheres that release alginate lyase [120]. These degradable scaffolds promoted the expansion of rat neural progenitor cells at higher levels than the nondegradable alginate scaffolds. Further work used this system to elucidate the role of the elastic modulus of these hydrogels in regulating stem cell behavior [121]. As elastic modulus increased, the proliferation rate of the stem cells decreased. A 2012 study confirmed that softer alginate hydrogels with an elastic modulus in the range of 0.1–1 kPa promoted neural growth and neurite extension from both rat and human neurons [122].

In 2009, the Kipke group generated a neural prosthesis probe coated with a mixture of alginate containing neural stem cells [123]. This coating helped to limit the cell death normally associated with the insertion and use of such probes for neural recordings. Further work from his group looked at how changing the composition of such alginate scaffolds affected the behavior of neural stem cells seeded inside [124]. Another study determined that applying an electric field of 1 Hz increased the viability of neural stem cells seeded inside of alginate microbeads [125]. Such alginate beads also support the culture and differentiation of mouse embryonic stem cells into neurons. The presence of retinoic acid enhances this effect [126]. Another group found similar results when using alginate microbeads to encapsulate mouse embryonic stem cells and differentiate them into neural lineages, finding that adding hyaluronic acid enhanced these effects [127]. These microbeads can encapsulate other primary cells and then combined with microbeads containing human embryonic stem cells to promote neuronal differentiation [128]. Other work indicated encapsulating human adipose-derived stem cells inside of alginate microbeads also encourages differentiation into neural phenotypes [129]. Another group used alginate scaffolds to deliver bFGF and EGF in a controlled fashion to rat neural progenitors seeded inside [129]. These scaffolds delivered the aforementioned factors for 21 days and promoted the proliferation, migration, and differentiation of the neural stem cells. A microfluidics-based approach generated tubular alginate hydrogels that enabled mouse neural stem cells to differentiate into aligned neural tissue as shown in Fig. 3 [130]. A recent set of studies showed how alginate can deliver stem cells for therapeutic applications in the damaged central nervous system. The first study transplanted rat neural stem cells into a rat model of spinal cord injury [131]. This combination reduced lesion size and

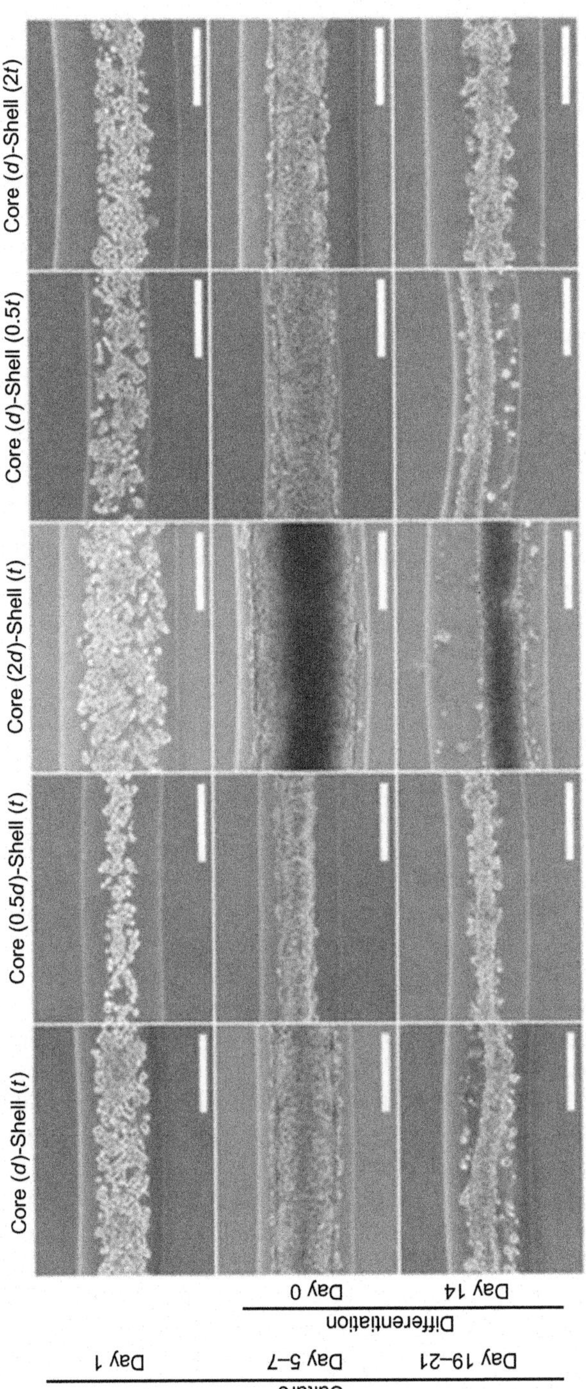

FIG. 3
Phase-contrast images of mouse neural stem cells cultured and differentiated in calcium alginate microtubes. All scale bars are 200 μm.

Reprinted with permission from Onoe H, Kato-Negishi M, Itou A, Takeuchi S. Differentiation induction of mouse neural stem cells in hydrogel tubular microenvironments with controlled tube dimensions. Adv Healthcare Mater 2016;5(9):1104–11.

inflammation while promoting neurological recovery. Another recent study seeded neural progenitors derived from both human embryonic stem cells and human-induced pluripotent stem cells inside of an alginate scaffold functionalized with the RGD peptide [132]. These cells then differentiated into mature retinal tissue, marking first time such scaffolds had been applied in that manner.

The ease of cross-linking alginate makes it an attractive biomaterial for applications in 3D bioprinting [133]. For example, microarrays of alginate beads encapsulating mouse embryonic stem cells were used to screen for factors that promoted neural differentiation [134]. Such arrays can also be generated using human neural stem cells and these arrays can be used for applications in screening drugs for toxicity [135]. Other applications include using microfluidics to print alginate-based stem cell niches that support the culture of human neural stem cells [136]. Overall, these examples demonstrate the versatility of alginate as a biomaterial for neural tissue engineering applications. The use of alginate as a bioink will be an interesting area for future work when engineering neural tissue using 3D printing technology.

2.2.3 Chitosan

Chitosan is another popular polysaccharide biomaterial used for tissue engineering applications [137]. It consists of a linear polysaccharide derived by deacetylating chitin, a biomolecule found in the shells of shrimp. This treatment process makes chitosan more biocompatible by increasing its solubility in water. Chitosan has a positive charge under normal physiological conditions, which enables it to bind to negatively charged surfaces. In addition to tissue engineering applications, chitosan can be used as a fining agent during the wine-making process to clear the wine before bottling. It can also be polymerized into hydrogels using different types of cross-linking reagents that improve its stability by creating covalent bonds between the individual chitosan molecules [138]. Using different cross-linking reagents enable the tuning the mechanical properties of chitosan scaffolds to function more effectively depending on the chosen application [139]. An interesting study looked at the ability of different chitosan formulations to support rat neural stem cell culture and found that chitosan films promoted more cell proliferation compared to hydrogel and conduit formulations [140].

Initial studies looked at the use of chitosan coatings to support neural stem cell culture [141,142]. The Tator lab working in collaboration with the Shoichet lab and Dr. Cindy Moorshead's group conducted an in vivo study where tubular chitosan scaffolds were used to deliver neural stem cells in the injured spinal cord [143–145]. These neural stem cells were derived from both the brain and spinal cord, with the brain-derived neural stem cells generating thicker tissue bridges than the spinal cord-derived neural stem cells. While the cells survived and differentiated in mature neural phenotypes, these engineered tissues did not promote functional recovery. Further work incorporated growth factor delivery alongside controlled release of antibodies against the Nogo receptor in conjunction with these stem cell seeded scaffolds, but such a combinatorial approach did not lead to an increase in functional recovery in a complete transection injury model [146]. More recent work from two separate groups demonstrated other combinations of chitosan scaffolds containing growth factors and stem cells promoted functional recovery in other, less severe preclinical models of spinal cord injury [147,148].

Chitosan can also be used for drug delivery applications. Chitosan nanocarriers can deliver NT-3 when incorporated into neurospheres where they promoted both cell proliferation and neuronal differentiation compared to unloaded control nanocarriers [149]. Follow-up work explored the effect of different NT-3 doses on neural stem cell behavior [150]. A modified version of chitosan was designed to deliver interferon gamma to neural stem cells as a way to promote neuronal differentiation [151]. Further work using the same type of chitosan scaffolds showed that a Young's elastic modulus of 3.5 kPa was optimal for promoting proliferation of neural stem cells [152]. Later work from the Shoichet group formulated a photopolymerizable version of chitosan that provided controlled release of interferon gamma to promote differentiation of neural stem cells into mature neurons and oligodendrocytes [153]. The Leipzig group followed up on this work by modulating the pore size of these chitosan gels, showing that large pores were more conducive to neural stem cell growth [154]. They demonstrated that the chitosan scaffolds that delivered growth factors would work in an in vivo setting [155]. Other work used chitosan scaffolds to deliver BDNF to mesenchymal stem cells seeded inside as it promoted an increase in cell number [156].

A group of researchers also evaluated the ability of a chitosan nerve guidance conduit to support the differentiation of bone marrow-derived mesenchymal stem cells into neuron-like cells and demonstrated these constructs could be used to treat peripheral nerve injury in preclinical testing [157]. Similar work demonstrated the feasibility of using adipose-derived stem cells with chitosan scaffolds for peripheral nerve repair as well where this combination promoted functional recovery in preclinical testing [158]. Other work showed that neural stem cells seeded inside of a chitosan scaffold and treated with NGF differentiated into neurons [159]. Another group used 3D chitosan scaffolds as a platform for real time imaging of fluorescently labelled neural stem cells seeded inside as they proliferated and differentiated [160]. Overall, the large set of studies detailed in this section show the versatility of chitosan and indicate its importance as natural biomaterial scaffold for engineering neural tissue, as it supports neural stem cell culture and can deliver growth factors in a controlled fashion.

2.2.4 Methylcellulose and nitrocellulose

Cellulose is a common linear polysaccharide found in both plants and bacteria that has many medical applications [161,162]. For biomedical applications, cellulose is often chemically modified to produce compounds such as methylcellulose and nitrocellulose, although some work has been published on the feasibility of using cellulose as a stem cell substrate [163]. For example, human adipose-derived stem cells cultured inside of microporous scaffolds derived from cellulose expressed neural markers, suggesting such materials are suitable for neural tissue engineering [164]. Cellulose scaffolds can be fabricated to display novel microtopography using a combination of photolithography and etching [165]. Methylcellulose is a derivative of cellulose that is easier to use for tissue engineering applications as it has better biocompatability. Work from Dr. Michelle LaPlaca's and Dr. Andres Garcia's groups at the Georgia Institute of Technology demonstrated that 3D methylcellulose scaffolds functionalized with laminin supported the culture of mouse neural stem cells and enhanced their migration within the scaffold [166]. In Section 2.3, a combination of methylcellulose and hyaluronan developed by the Shoichet lab will be discussed extensively as it has been used for a variety of neural tissue engineering applications.

2.2.5 Dextran

Unlike some of the other polysaccharides discussed in this chapter, dextran possesses a highly branched structure as opposed to being a linear sugar molecule. In terms of sourcing, bacteria polymerize sucrose molecules into dextran chains. It possesses highly reactive hydroxyl groups, making it easy to modify into forms that self-assemble into hydrogels and nanocarriers for tissue engineering applications [167]. It is also commonly used to encapsulate different types of molecules for applications in drug delivery [168]. Intriguingly, dextran can act as an anticoagulant, which may prove beneficial when engineering tissue to replace those lost to traumatic injury [169]. It also can be used to trace the tracts of the nervous system [170]. These aforementioned properties make it an attractive biomaterial for developing replacement neural tissue.

In one example of neural tissue engineering, dextran hydrogels functionalized with laminin peptides were produced using cryogelation and then seeded with human cord blood-derived stem cells that then differentiated into neuron-like cells [171]. These scaffolds were then implanted into the rat brain where they attracted host cells to migrate into the scaffolds, demonstrating their neuroattractive properties. However, they did not reduce the inflammation present post injury. A recent study demonstrated 3D dextran scaffolds could support the culture of mouse embryonic stem cells while reducing the need for passaging [172]. This study suggests that such materials applied to neural tissue engineering as well. Chapter 6 gives examples of how dextran can be used in combination with synthetic biomaterials for engineering the neural stem cell niche.

2.2.6 Hyaluronan/hyaluronic acid

The nervous system contains significant amounts of hyaluronan, which is a linear polysaccharide often referred to as hyaluronic acid [173]. It can also be derived from plants or other animals as well. Hyaluronan can bind chondroitin sulfate proteoglycans, and it plays major role in controlling cell migration, indicating its bioactive nature. In 2009, Wang and Spector fabricated 3D hyaluronan scaffolds functionalized with collagen that supported the culture and differentiation of rat neural stem cells [174]. A different study used 3D hyaluronic acid scaffolds in combination with other factors (antibody against the Nogo receptor and poly(L-lysine) (PLL) to determine their effect on the behavior of rat neural stem cells [175]. They found incorporating the antibody against the Nogo receptor did not affect their behavior, but PLL did inhibit differentiation. The mechanical properties of such 3D hyaluronic acid scaffolds can be modified and these properties influence stem cell differentiation [176]. They observed stiffer gels tended to differentiate of neural progenitor cells into astrocytes. In terms of in vivo work, using 3D hyaluronic acid scaffolds to deliver human neural stem cells into the mouse brain increased their rate of survival [177].

Another recent study optimized the concentrations of three different peptide sequences (RGD, YIGSR, IKVAV) inside of 3D hyaluronic acid for supporting the culture and differentiation of human-induced pluripotent stem cells into neural tissue [178]. The ability of these scaffolds to promote optimal levels of proliferation and differentiation of these cells was highly dependent on the concentration of the peptides incorporated. Such scaffolds were then used to deliver these neural progenitors derived from human-induced pluripotent stem cells into a preclinical model of stroke, where they promoted differentiation of the cells seeded inside but did not increase survival [179]. Hyaluronic acid is popular biomaterial to use in combination with others to generate multifunctional scaffolds as discussed later in this chapter and in Chapter 6.

2.3 MIXTURES OF THE DIFFERENT NATURAL EXTRACELLULAR MATRIX COMPONENTS AS SCAFFOLDS FOR CONTROLLING STEM CELL BEHAVIOR FOR NEURAL TISSUE ENGINEERING APPLICATIONS

This section will examine selected relevant examples of scaffolds where two or more of the biomolecules described above have been combined for applications in generating neural tissue from stem cells. As detailed in Sections 2.1.1 and 2.1.2, collagen and fibrin are two commonly used biomaterials for neural tissue engineering. Thus, many groups have performed studies where they have added more bioactive components to these protein-based scaffolds as a way to enhance their properties. For example, a scaffold combining fibronectin and collagen was successfully used to deliver mouse-derived fetal neural stem cells into a preclinical model of traumatic brain injury [83]. These cells survived and could migrate in the injured brain. Another study combined hyaluronic acid with collagen to deliver bFGF to increase the survival of neural stem cells. Such 3D hyaluronic acid scaffolds can be combined with laminin-derived peptides and BDGF to deliver human mesenchymal stem cells in the injured rat spinal cord injury site where this combination promoted an increase in functional recovery [180]. A 2016 study combined hyaluronic acid with collagen to generate a scaffold that was then loaded with rat neural stem cells and bFGF [181]. This combination of bioactive materials and bFGF quickly generated functional neurons, and similar differentiation was observed when this engineered tissue was transplanted into preclinical model of traumatic brain injury as seen in Fig. 4. This study built upon previous work demonstrating that such blends of collagen and hyaluronic acid support neuronal differentiation of adult neural stem cells in vitro [182], and such combinations could be used for repairing facial nerves in a rabbit injury model [183]. Also, these blends can be formulated to vary the stiffness of such collagen scaffolds, which can in turn modulate stem cell differentiation [184]. Another group printed 3D collagen/fibrin scaffolds containing vascular endothelial growth factor (VEGF) that supported the culture of murine neural stem cells, further extending the utility of this material [185]. Other work has combined collagen with laminin-derived peptides for supporting the survival of neural

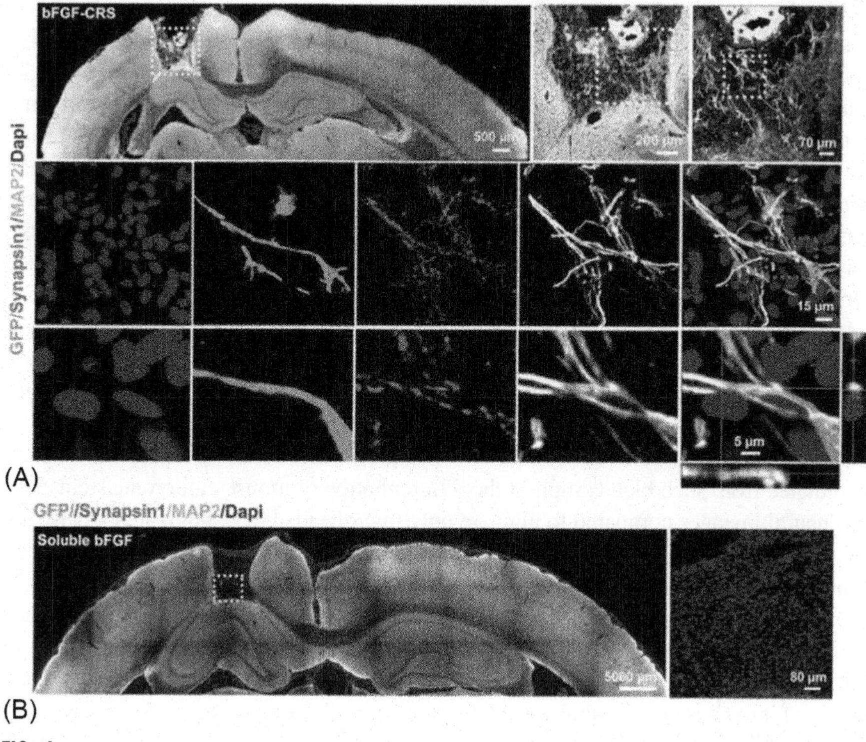

FIG. 4

A combination scaffolds consisting of hyaularonate and collagen can deliver bFGF while promoting the differentiation of transplanted neural stem cells (NSCs) into mature neurons. (A) Four weeks after the transplantation, the GFP+ NSCs were observed in the lesion area of the bFGF-CRS group. The tetra-label immunofluorenscence results suggested that the transplanted NSCs *(green in the electronic version and gray in the printed version)* could differentiate into mature neurons (MAP2+ *white*). The synapsin1+particles *(red in the electronic version and gray in the printed version)* adhered to the GFP-positive dendrites, indicating that the transplanted NSCs had differentiated into neurons and could form synapses input. (B) Four weeks after the transplantation, few GFP+ and MAP2+cells were observed in the brain injured area of the soluble bFGF group.

Reprinted with permission from Duan H, Li X, Wang C, Hao P, Song W, Li M, et al. Functional hyaluronate collagen scaffolds induce NSCs differentiation into functional neurons in repairing the traumatic brain injury. Acta Biomater 2016;45:182–95.

stem cells in the short term over 24 and 48h [186]. In a similar work, two studies used Matrigel to deliver both neural- and adipose-derived stem cells into canine spinal cord injury models [187,188], but the sourcing of Matrigel makes it hard to translate such results for clinical applications as it is derived from mouse cancer cells.

Fibrin has also been combined with other biological molecules to produce multifunctional scaffolds for producing neural tissue from stem cells. Combining fibrin and hyaluronic acid, as well as the aforementioned mixture with laminin, resulted in scaffolds that promoted the differentiation of hematopoietic tissue-derived adult stem cells into neural progenitors more effectively than fibrin only or 2D fibronectin coated surfaces [88]. Another study used a 3D fibrin scaffolds that incorporated laminin and fibronectin to direct rat neural stem cells to differentiate into motor neurons as indicated by choline acetyltransferase staining [189]. Another intriguing combination of materials is the use of fibrin and alginate to produce scaffolds for repairing peripheral nerve injuries as the alginate enhances the mechanical stability of these constructs. In one example, these blended scaffolds were used to deliver adipose-derived mesenchymal stem cells seeded inside of a nerve guidance conduit to treat a 10-mm sciatic nerve injury in rats, leading to more effective nerve regeneration [190]. Further in vitro work generated a nanostructured version of this composite scaffold to optimize the behavior of the adipose-derived mesenchymal stem cells seeded inside, including cell clustering and proliferation [191].

Alginate can be modified to contain both hyaluronic acid and fibronectin. Microbeads formulated from such blends support the differentiation of mouse embryonic stem cells into neural lineages compared to alginate only microbeads [127]. Alginate can also be mixed with gelatin to make bioactive scaffolds. Microbeads made from such formulations support the expansion of neural stem cells in format compatible bioreactors [192]. Interesting, alternating layers of gelatin and alginate can also encapsulate individual neural stem cells [193]. These layers could be loaded with insulin-like growth factor 1 (IGF-1), which enhanced stem cell proliferation. This work provides an interesting strategy for expansion of neural stem cells using growth factors.

Other blends include mixing chitin and alginate to make microfiber scaffolds that supported highly efficient differentiation of both human embryonic and induced pluripotent stem cell lines into neurons [194]. Additionally different mixtures of alginate with other polysaccharides serve as effective bioinks for 3D printing functional mini-tissues [195]. As with collagen, a recent study showed that silk scaffolds containing 3%–6% hyaluronic acid supported the culture of primary neural cells, indicating such a scaffold would be suitable for neural tissue engineering applications [196].

A large body of work done by a group of researchers at the University of Toronto—led by prominent biomaterials scientist and University Professor, Dr. Molly Shoichet, has validated a novel biomaterial that combines hyaluronan and methylcellulose, which is abbreviated as HAMC. In a pair of studies, they demonstrated how this material could deliver controlled release of various proteins to the intrathecal space found in the spinal cord of rats [197–199]. This material was then adapted for delivering controlled release of the small molecule, nimodipine, by generating particles containing this molecule instead of hydrogels [200]. Follow-up work used this scaffold to deliver adult brain-derived neural stem cells to the injured spinal cord [201]. These engineered tissues reduced cavitation and increased survival of the transplanted cells. While the initial studies were in the context of spinal cord injury, they later used an injectable version of this combination of hyaluronan and methylcellulose seeded with retinal stem-progenitor cells to deliver these cells into the sub-retinal space in mice as seen in Fig. 5 [202]. The hydrogel ensured a more even distribution of cells in vivo

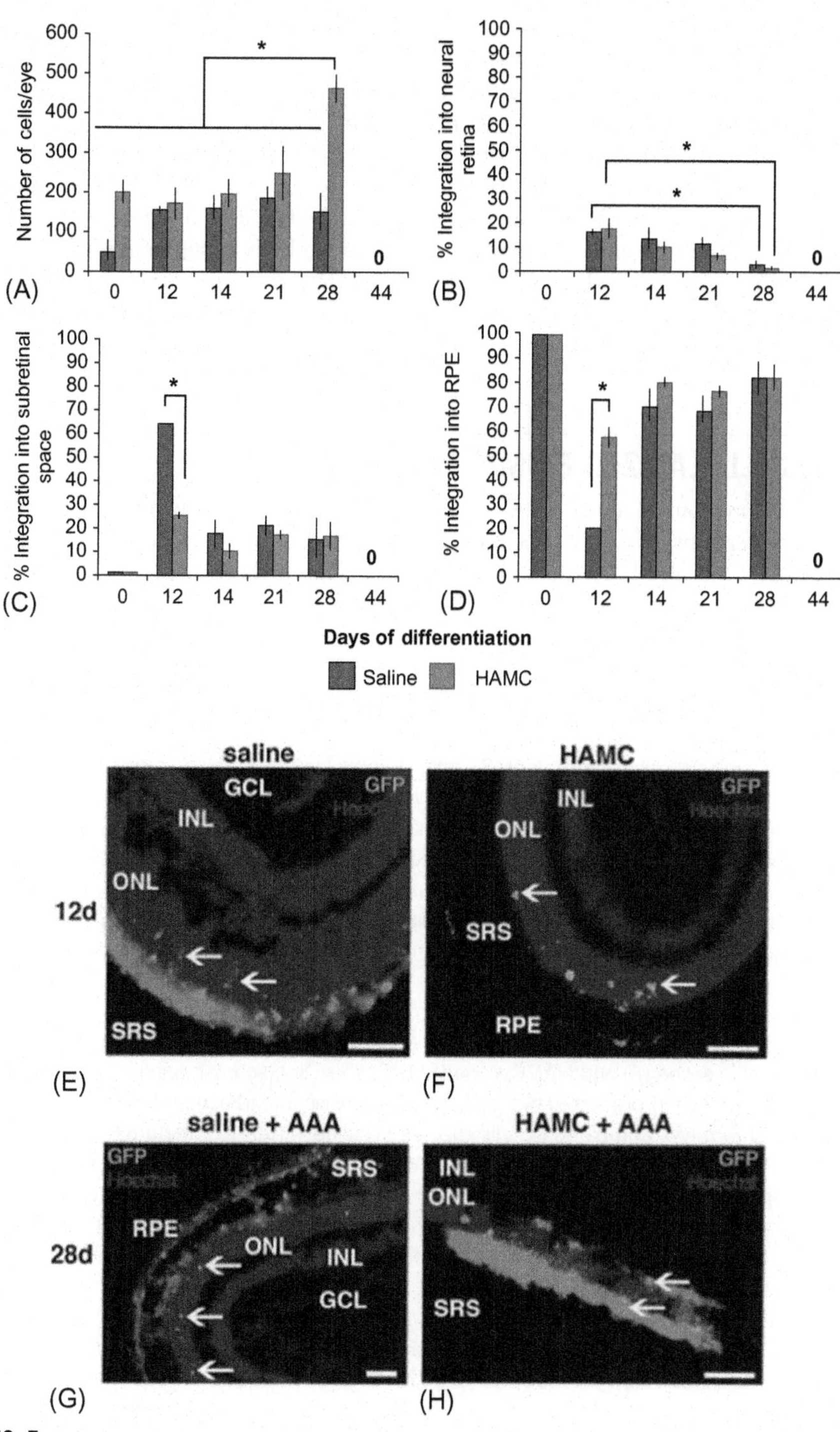

FIG. 5

See the legend on next page

compared to delivery using a saline solution. Other work has used this material to deliver drugs improve recovery after stroke [203,204]. Thus, this combinatorial material referred to as HAMC, reflecting both hyaluronan and methylcellulose, can be applied to diverse applications in repairing the damaging nervous system.

Overall, the set of studies presented here show how enhanced biological effects can be achieved by combining the effects of multiple bioactive molecules and that such strategies can lead to improvements in function when used as engineered scaffolds for controlling stem cell behavior. While the work discussed in this section is not an exhaustive list of all combinatorial approaches, it does provide an effective overview on how such combinations can be applied to generate bioactive scaffolds for engineering neural tissue.

3 DECELLULARIZED BIOMATERIALS

While the previous sections focused on homogenous naturally derived biomaterial scaffolds, nerve tissue can be processed using decellularization to generate scaffolds as reviewed recently [205,206]. These methods often involved treating the tissues with detergents to lyse the cells while leaving the extracellular matrix and its structure intact. The method of treatment will determine the resulting properties of the

FIG. 5

Hyaluronan and methylcellulose (HAMC), an injectable hydrogel matrix, improves retinal stem cell (RSC)-derived rod photoreceptor survival after transplantation. (A–D) The absolute numbers of cells surviving in the retina of adult mice 3 weeks posttransplantation in saline versus HAMC vehicle (A). HAMC encourages the posttransplant survival of RSC-derived rods compared to saline as the maturity of the rods increases. Quantification was performed on the percentage of integrating cells (as a fraction of total cells counted in the whole eye) (B) in the neural retina, (C) in the subretinal space, and (D) adherent to the RPE layer. Immature RSC-derived rods show an ability to integrate into host neural retina independent of delivery vehicle. Mean ± SEM of $n=4$–8 independent transplants are plotted; $*p<0.05$. (E)–(H) Inverted epifluorescent image of immature RSC-derived rods transplanted in saline (E) and HAMC alone (F) both show sparse integration into the ONL and extension of short processes but fail to show extensive integration and mature rod morphology *(arrows)*. Compared to cells injected in saline + AAA (G), mature RSC-derived rods in HAMC (28-day RSC-derived rods in HAMC + AAA) (H) show patches of extensive integration into the ONL and mature rod morphology, including the extension of processes toward the outer plexiform layer and outer segments *(arrows)*, compared to cells injected in saline. Scale bars represent 50 μm. *GCL*, ganglion cell layer; *INL*, inner nuclear layer; *ONL*, outer nuclear layer; *RPE*, retinal pigment epithelium layer; *SRS*, subretinal space.

Reprinted with permission from Ballios BG, Cooke MJ, Donaldson L, Coles BL, Morshead CM, van der Kooy D, et al. A hyaluronan-based injectable hydrogel improves the survival and integration of stem cell progeny following transplantation. Stem Cell Rep 2015;4(6):1031–45; Taken from https://www.ncbi.nlm.nih.gov/pmc/articles/PMC4471829/.

scaffold, making it an important consideration when preparing such scaffolds. Most of these approaches for decellularizing neural tissue have focused on the peripheral nervous system as this system has a higher regenerative capacity in comparison with the central nervous system tissue. The gold standard for repairing peripheral nerve damage is the use of autografts although recent research has begun to explore the use of decellularized nerve tissue as an attractive alternative [207]. Several examples, including work from Dr. Christine Schmidt's group at the University of Florida, demonstrated the potential of acellular nerve grafts produced by decellularization to promote regeneration in preclinical models of nerve injuries [208–210].

Recent efforts have supplemented these extracellular matrix-derived scaffolds with stem cells to promote further regeneration in the peripheral nervous system. For example, seeding acellular scaffolds with bone marrow-derived stromal cells encapsulated in fibrin glue enhanced recovery of function post injury compared to the acellular scaffolds alone [211]. Another study evaluated the effect of seeding rat-derived hair-follicle neural crest stem cells into acellular nerve graft and then transplanting these constructs in a peripheral nerve injury model in beagles [212]. Finally, other research used an acellular scaffold derived from spinal cord tissue seeded with mesenchymal stem cells to promote both long distance axonal regeneration and functional recovery in a rat model of spinal cord injury [213]. These studies show promising results and this area remains understudied, making it an excellent area for future work as detailed in the next section.

4 FUTURE DIRECTIONS

This chapter has detailed the significant body of research exploring how a variety of naturally derived biomaterials can be applied to neural tissue engineering using stem cells. While these bioactive scaffolds can successfully differentiate stem cells into neural tissue in vitro and in preclinical testing, significant amounts of work remain to successfully translate these technologies for clinical applications. Producing these scaffolds in a reproducible and high throughput fashion requires further investigation and these processes may benefit from the advent of 3D bioprinting as discussed in the Chapter 8. Other barriers to clinical translation include the high level of safety data necessary for gaining approval to transplant cellular therapies (as opposed to only biomaterial scaffolds). As mentioned in the last section, the use of decellularized extracellular matrix has yet to be explored fully for applications using tissues derived from the central nervous system. Furthermore, there are several naturally derived proteins and polysaccharides that have not been investigated for their compatibility with stem cells and thus remain a target for future work. For example, the use of hydrogels made from chondroitin sulfates can emulate the neural stem cell niche [214] and such materials require in-depth study to determine their potential for tissue engineering applications. The work detailed in this chapter provides an excellent place to begin for fully reviewing the potential of engineering neural tissue using stem cells.

ADDITIONAL RESOURCES

LINKS TO WEB RESOURCES ABOUT BIOMATERIALS

Society for Biomaterials: https://www.biomaterials.org/

Canadian Society for Biomaterials: https://biomaterials.ca/

Tissue Engineering and Regenerative Medicine International Society: https://www.termis.org/

Biomedical Engineering Society: http://www.bmes.org/

Nature's website on Biomaterial Research: http://www.nature.com/subjects/biomaterials

Biomaterials—the journal: http://www.journals.elsevier.com/biomaterials/

Tissue Engineering—the journal: http://www.liebertpub.com/overview/tissue-engineering-parts-a-b-and-c/595/

TEXTBOOKS ABOUT BIOMATERIALS FOR FURTHER READING

Biomaterials Science: An Introduction to Materials in Medicine (Third Edition). Edited by Buddy Ratner, Allan S. Hoffman, Fredrick J. Schoen, and Jack E. Lemons. 3rd edition. Massachusetts: Elsevier. 2013.

Temenoff, J.S. and Mikos, A.G. Biomaterials: The Intersection of Biology and Materials Science. 1st edition. London: Pearson Science. 2008.

Williams, D.F. Essential Biomaterial Science (Cambridge Texts in Biomedical Engineering). 1st edition. Cambridge: Cambridge University Press 2014.

TEXTBOOKS ABOUT TISSUE ENGINEERING FOR FURTHER READING

Saltzman, M. Tissue Engineering: Engineering Principles for the Design of Replacement Organs and Tissues. Oxford University Press. 1st edition. 2004.

Principles of Tissue Engineering. Edited by Robert Lanza, Robert Langer, and Joseph Vacanti. Academic Press. Fourth edition. 2013.

REFERENCES

[1] Ratner BD, Hoffman AS, Schoen FJ, Lemons JE. Biomaterials science: an introduction to materials in medicine. Cambridge, MA: Academic Press; 2004.

[2] Khaing ZZ, Schmidt CE. Advances in natural biomaterials for nerve tissue repair. Neurosci Lett 2012;519(2):103–14.

[3] Omstead DR, Baird LG, Christenson L, Moulin GD, Tubo R, Maxted DD, et al. Voluntary guidance for the development of tissue-engineered products. Tissue Eng 1998;4(3):239–66.

[4] Hussein KH, Park K-M, Kang K-S, Woo H-M. Biocompatibility evaluation of tissue-engineered decellularized scaffolds for biomedical application. Mater Sci Eng C 2016;67:766–78.

[5] Gould LJ. Topical collagen-based biomaterials for chronic wounds: rationale and clinical application. Adv Wound Care 2016;5(1):19–31.

[6] Mehta-Ambalal SR. Neocollagenesis and neoelastinogenesis: from the laboratory to the clinic. J Cutan Aesthet Surg 2016;9(3):145.

[7] Hapach LA, VanderBurgh JA, Miller JP, Reinhart-King CA. Manipulation of in vitro collagen matrix architecture for scaffolds of improved physiological relevance. Phys Biol 2015;12(6):061002.

[8] Yannas I, Tzeranis D, So P. Surface biology of collagen scaffold explains blocking of wound contraction and regeneration of skin and peripheral nerves. Biomed Mater 2015;11(1):014106.

[9] O'Connor SM, Stenger DA, Shaffer KM, Maric D, Barker JL, Ma W. Primary neural precursor cell expansion, differentiation and cytosolic Ca 2+ response in three-dimensional collagen gel. J Neurosci Methods 2000;102(2):187–95.

[10] Ma W, Fitzgerald W, Liu Q-Y, O'shaughnessy T, Maric D, Lin H, et al. CNS stem and progenitor cell differentiation into functional neuronal circuits in three-dimensional collagen gels. Exp Neurol 2004;190(2):276–88.

[11] Itoh T, Satou T, Dote K, Hashimoto S, Ito H. Effect of basic fibroblast growth factor on cultured rat neural stem cell in three-dimensional collagen gel. Neurol Res 2005;27(4):429–32.

[12] Lin HJ, O'Shaughnessy TJ, Kelly J, Ma W. Neural stem cell differentiation in a cell–collagen–bioreactor culture system. Dev Brain Res 2004;153(2):163–73.

[13] Watanabe K, Nakamura M, Okano H, Toyama Y. Establishment of three-dimensional culture of neural stem/progenitor cells in collagen type-1 gel. Restor Neurol Neurosci 2007;25(2):109–17.

[14] Huang F, Shen Q, Zhao J. Growth and differentiation of neural stem cells in a three-dimensional collagen gel scaffold. Neural Regen Res 2013;8(4):313.

[15] Ge D, Song K, Guan S, Qi Y, Guan B, Li W, et al. Culture and differentiation of rat neural stem/progenitor cells in a three-dimensional collagen scaffold. Appl Biochem Biotechnol 2013;170(2):406–19.

[16] Führmann T, Hillen LM, Montzka K, Wöltje M, Brook GA. Cell–cell interactions of human neural progenitor-derived astrocytes within a microstructured 3D-scaffold. Biomaterials 2010;31(30):7705–15.

[17] Bercu MM, Arien-Zakay H, Stoler D, Lecht S, Lelkes PI, Samuel S, et al. Enhanced survival and neurite network formation of human umbilical cord blood neuronal progenitors in three-dimensional collagen constructs. J Mol Neurosci 2013;51(2):249–61.

[18] Mori H, Takahashi A, Horimoto A, Hara M. Migration of glial cells differentiated from neurosphere-forming neural stem/progenitor cells depends on the stiffness of the chemically cross-linked collagen gel substrate. Neurosci Lett 2013;555:1–6.

[19] Song K, Ge D, Guan S, Sun C, Ma X, Liu T. Mass transfer analysis of growth and substance metabolism of NSCs cultured in collagen-based scaffold in vitro. Appl Biochem Biotechnol 2014;174(6):2114–30.

[20] Hughes CS, Postovit LM, Lajoie GA. Matrigel: a complex protein mixture required for optimal growth of cell culture. Proteomics 2010;10(9):1886–90.

[21] Chen SS, Revoltella RP, Zimmerberg J, Margolis L. Differentiation of rhesus monkey embryonic stem cells in three-dimensional collagen matrix. Methods Mol Biol 2006;330:431–43.

[22] Kothapalli CR, Kamm RD. 3D matrix microenvironment for targeted differentiation of embryonic stem cells into neural and glial lineages. Biomaterials 2013;34(25):5995–6007.

[23] Khayyatan F, Nemati S, Kiani S, Emami SH, Baharvand H. Behaviour of human induced pluripotent stem cell-derived neural progenitors on collagen scaffolds varied in freezing temperature and laminin concentration. Cell J (Yakhteh) 2014;16(1):53.

[24] Macri-Pellizzeri L, Pelacho B, Sancho A, Iglesias-Garcia O, Simon-Yarza AM, Soriano-Navarro M, et al. Substrate stiffness and composition specifically direct differentiation of induced pluripotent stem cells. Tissue Eng Part A 2015;21(9–10):1633–41.

[25] Egawa EY, Kato K, Hiraoka M, Nakaji-Hirabayashi T, Iwata H. Enhanced proliferation of neural stem cells in a collagen hydrogel incorporating engineered epidermal growth factor. Biomaterials 2011;32(21):4737–43.

[26] Ma F, Xiao Z, Chen B, Hou X, Han J, Zhao Y, et al. Accelerating proliferation of neural stem/progenitor cells in collagen sponges immobilized with engineered basic fibroblast growth factor for nervous system tissue engineering. Biomacromolecules 2014;15(3):1062–8.

[27] Huang F, Wu Y, Wang H, Chang J, Ma G, Yin Z. Effect of controlled release of brain-derived neurotrophic factor and neurotrophin-3 from collagen gel on neural stem cells. NeuroReport 2016;27(2):116–23.

[28] Nakaji-Hirabayashi T, Kato K, Iwata H. Improvement of neural stem cell survival in collagen hydrogels by incorporating laminin-derived cell adhesive polypeptides. Bioconjug Chem 2012;23(2):212–21.

[29] Yang Z, Qiao H, Sun Z, Li X. Effect of BDNF–plasma–collagen matrix controlled delivery system on the behavior of adult rats neural stem cells. J Biomed Mater Res A 2013;101(2):599–606.

[30] Huang F, Yin Z, Wu D, Hao J. Effects of controlled release of brain-derived neurotrophic factor from collagen gel on rat neural stem cells. NeuroReport 2013;24(3):101–7.

[31] Shamloo A, Heibatollahi M, Mofrad MR. Directional migration and differentiation of neural stem cells within three-dimensional microenvironments. Integr Biol 2015;7(3):335–44.

[32] Chen Z, Lu X-CM, Shear DA, Dave JR, Davis AR, Evangelista CA, et al. Synergism of human amnion-derived multipotent progenitor (AMP) cells and a collagen scaffold in promoting brain wound recovery: pre-clinical studies in an experimental model of penetrating ballistic-like brain injury. Brain Res 2011;1368:71–81.

[33] Mahmood A, Wu H, Qu C, Mahmood S, Xiong Y, Kaplan DL, et al. Suppression of neurocan and enhancement of axonal density in rats after treatment of traumatic brain injury with scaffolds impregnated with bone marrow stromal cells: laboratory investigation. J Neurosurg 2014;120(5):1147–55.

[34] Elias PZ, Spector M. Implantation of a collagen scaffold seeded with adult rat hippocampal progenitors in a rat model of penetrating brain injury. J Neurosci Methods 2012;209(1):199–211.

[35] Nie X, Zhang Y-J, Tian W-D, Jiang M, Dong R, Chen J-W, et al. Improvement of peripheral nerve regeneration by a tissue-engineered nerve filled with ectomesenchymal stem cells. Int J Oral Maxillofac Surg 2007;36(1):32–8.

[36] Georgiou M, Golding JP, Loughlin AJ, Kingham PJ, Phillips JB. Engineered neural tissue with aligned, differentiated adipose-derived stem cells promotes peripheral nerve regeneration across a critical sized defect in rat sciatic nerve. Biomaterials 2015;37:242–51.

[37] Matsuse D, Kitada M, Ogura F, Wakao S, Kohama M, J-i K, et al. Combined transplantation of bone marrow stromal cell-derived neural progenitor cells with a collagen sponge and basic fibroblast growth factor releasing microspheres enhances recovery after cerebral ischemia in rats. Tissue Eng A 2011;17(15–16):1993–2004.

[38] Yu H, Cao B, Feng M, Zhou Q, Sun X, Wu S, et al. Combinated transplantation of neural stem cells and collagen type I promote functional recovery after cerebral ischemia in rats. Anat Rec 2010;293(5):911–7.

[39] Cholas RH, Hsu HP, Spector M. The reparative response to cross-linked collagen-based scaffolds in a rat spinal cord gap model. Biomaterials 2012;33(7):2050–9.

[40] Li X, Xiao Z, Han J, Chen L, Xiao H, Ma F, et al. Promotion of neuronal differentiation of neural progenitor cells by using EGFR antibody functionalized collagen scaffolds for spinal cord injury repair. Biomaterials 2013;34(21):5107–16.

[41] Yuan N, Tian W, Sun L, Yuan R, Tao J, Chen D. Neural stem cell transplantation in a double-layer collagen membrane with unequal pore sizes for spinal cord injury repair. Neural Regen Res 2014;9(10):1014–9.

[42] Sugai K, Nishimura S, Kato-Negishi M, Onoe H, Iwanaga S, Toyama Y, et al. Neural stem/progenitor cell-laden microfibers promote transplant survival in a mouse transected spinal cord injury model. J Neurosci Res 2015;93(12):1826–38.

[43] Ishikawa M, Ohnishi H, Skerleva D, Sakamoto T, Yamamoto N, Hotta A, et al. Transplantation of neurons derived from human iPS cells cultured on collagen matrix into guinea-pig cochleae. J Tissue Eng Regen Med 2015;1–13.

[44] Liu T, Teng WK, Chan BP, Chew SY. Photochemical crosslinked electrospun collagen nanofibers: synthesis, characterization and neural stem cell interactions. J Biomed Mater Res A 2010;95A(1):276–82.

[45] Wang Y, Yao M, Zhou J, Zheng W, Zhou C, Dong D, et al. The promotion of neural progenitor cells proliferation by aligned and randomly oriented collagen nanofibers through β1 integrin/MAPK signaling pathway. Biomaterials 2011;32(28):6737–44.

[46] Winter CC, Katiyar KS, Hernandez NS, Song YJ, Struzyna LA, Harris JP, et al. Transplantable living scaffolds comprised of micro-tissue engineered aligned astrocyte networks to facilitate central nervous system regeneration. Acta Biomater 2016;38:44–58.

[47] Yin Y, Huang P, Han Z, Wei G, Zhou C, Wen J, et al. Collagen nanofibers facilitated presynaptic maturation in differentiated neurons from spinal-cord-derived neural stem cells through MAPK/ERK1/2-Synapsin I signaling pathway. Biomacromolecules 2014;15(7):2449–60.

[48] Li X, Liang H, Sun J, Zhuang Y, Xu B, Dai J. Electrospun collagen fibers with spatial patterning of SDF1α for the guidance of neural stem cells. Adv Healthcare Mater 2015;4(12):1869–76.

[49] Yao L, Phan F, Li Y. Collagen microsphere serving as a cell carrier supports oligodendrocyte progenitor cell growth and differentiation for neurite myelination in vitro. Stem Cell Res Therap 2013;4(5):1.

[50] Monsonego A, Mizrahi T, Eitan S, Moalem G, Bardos H, Adany R, et al. Factor XIIIa as a nerve-associated transglutaminase. FASEB J 1998;12(12):1163–71.

[51] Ahmed TA, Dare EV, Hincke M. Fibrin: a versatile scaffold for tissue engineering applications. Tissue Eng Part B Rev 2008;14(2):199–215.

[52] Rajangam T, An SS. Fibrinogen and fibrin based micro and nano scaffolds incorporated with drugs, proteins, cells and genes for therapeutic biomedical applications. Int J Nanomedicine 2013;8:3641–62.

[53] Schek RM, Michalek AJ, Iatridis JC. Genipin-crosslinked fibrin hydrogels as a potential adhesive to augment intervertebral disc annulus repair. Eur Cell Mater 2011;21:373–83.

[54] Likhitpanichkul M, Dreischarf M, Illien-Junger S, Walter BA, Nukaga T, Long RG, et al. Fibrin-genipin adhesive hydrogel for annulus fibrosus repair: performance

evaluation with large animal organ culture, in situ biomechanics, and in vivo degradation tests. Eur Cell Mater 2014;28:25–37. discussion -8.
[55] Willerth SM, Arendas KJ, Gottlieb DI, Sakiyama-Elbert SE. Optimization of fibrin scaffolds for differentiation of murine embryonic stem cells into neural lineage cells. Biomaterials 2006;27(36):5990–6003.
[56] Lorentz KM, Kontos S, Frey P, Hubbell JA. Engineered aprotinin for improved stability of fibrin biomaterials. Biomaterials 2011;32(2):430–8.
[57] Coffin ST, Gaudette GR. Aprotinin extends mechanical integrity time of cell-seeded fibrin sutures. J Biomed Mater Res A 2016;104(9):2271–9.
[58] Sakiyama SE, Schense JC, Hubbell JA. Incorporation of heparin-binding peptides into fibrin gels enhances neurite extension: an example of designer matrices in tissue engineering. FASEB J 1999;13(15):2214–24.
[59] Sakiyama-Elbert SE, Hubbell JA. Controlled release of nerve growth factor from a heparin-containing fibrin-based cell ingrowth matrix. J Control Release 2000;69(1):149–58.
[60] Sakiyama-Elbert SE, Hubbell JA. Development of fibrin derivatives for controlled release of heparin-binding growth factors. J Control Release 2000;65(3):389–402.
[61] Willerth SM, Faxel TE, Gottlieb DI, Sakiyama-Elbert SE. The effects of soluble growth factors on embryonic stem cell differentiation inside of fibrin scaffolds. Stem Cells 2007;25(9):2235–44.
[62] Willerth SM, Rader A, Sakiyama-Elbert SE. The effect of controlled growth factor delivery on embryonic stem cell differentiation inside fibrin scaffolds. Stem Cell Res 2008;1(3):205–18.
[63] Johnson PJ, Tatara A, Shiu A, Sakiyama-Elbert SE. Controlled release of neurotrophin-3 and platelet derived growth factor from fibrin scaffolds containing neural progenitor cells enhances survival and differentiation into neurons in a subacute model of SCI. Cell Transplant 2010;19(1):89.
[64] Johnson PJ, Tatara A, McCreedy DA, Shiu A, Sakiyama-Elbert SE. Tissue-engineered fibrin scaffolds containing neural progenitors enhance functional recovery in a subacute model of SCI. Soft Matter 2010;6(20):5127–37.
[65] McCreedy DA, Wilems TS, Xu H, Butts JC, Brown CR, Smith AW, et al. Survival, differentiation, and migration of high-purity mouse embryonic stem cell-derived progenitor motor neurons in fibrin scaffolds after sub-acute spinal cord injury. Biomater Sci 2014;2(11):1672–82.
[66] Wilems TS, Pardieck J, Iyer N, Sakiyama-Elbert SE. Combination therapy of stem cell derived neural progenitors and drug delivery of anti-inhibitory molecules for spinal cord injury. Acta Biomater 2015;28:23–32.
[67] Lu P, Wang Y, Graham L, McHale K, Gao M, Wu D, et al. Long-distance growth and connectivity of neural stem cells after severe spinal cord injury. Cell 2012;150(6):1264–73.
[68] Lu P, Graham L, Wang Y, Wu D, Tuszynski M. Promotion of survival and differentiation of neural stem cells with fibrin and growth factor cocktails after severe spinal cord injury. J Vis Exp 2014;89:e50641.
[69] Sharp KG, Yee KM, Steward O. A re-assessment of long distance growth and connectivity of neural stem cells after severe spinal cord injury. Exp Neurol 2014;257:186–204.
[70] Lu P, Woodruff G, Wang Y, Graham L, Hunt M, Wu D, et al. Long-distance axonal growth from human induced pluripotent stem cells after spinal cord injury. Neuron 2014;83(4):789–96.

[71] Montgomery A, Wong A, Gabers N, Willerth SM. Engineering personalized neural tissue by combining induced pluripotent stem cells with fibrin scaffolds. Biomater Sci 2015;3(2):401–13.

[72] Robinson M, Yau S, Sun L, Gabers N, Bibault E, Christie BR, et al. Optimizing differentiation protocols for producing dopaminergic neurons from human induced pluripotent stem cells for tissue engineering applications. Biomark Insights 2015;1(Suppl):61–70.

[73] Edgar JM, Robinson M, Willerth SM. Fibrin hydrogels induce mixed dorsal/ventral spinal neuron identities during differentiation of human induced pluripotent stem cells. Acta Biomater 2017;237–45.

[74] Shakhbazau A, Petyovka N, Kosmacheva S, Potapnev M. Neurogenic induction of human mesenchymal stem cells in fibrin 3D matrix. Bull Exp Biol Med 2011;150(4):547–50.

[75] Asmani MN, Ai J, Amoabediny G, Noroozi A, Azami M, Ebrahimi-Barough S, et al. Three-dimensional culture of differentiated endometrial stromal cells to oligodendrocyte progenitor cells (OPCs) in fibrin hydrogel. Cell Biol Int 2013;37(12):1340–9.

[76] Tara S, Krishnan LK. Bioengineered fibrin-based niche to direct outgrowth of circulating progenitors into neuron-like cells for potential use in cellular therapy. J Neural Eng 2015;12(3):036011.

[77] Bhang SH, Lee YE, Cho S-W, Shim J-W, Lee S-H, Choi CY, et al. Basic fibroblast growth factor promotes bone marrow stromal cell transplantation-mediated neural regeneration in traumatic brain injury. Biochem Biophys Res Commun 2007;359(1):40–5.

[78] Chen S-J, Chang C-M, Tsai S-K, Chang Y-L, Chou S-J, Huang S-S, et al. Functional improvement of focal cerebral ischemia injury by subdural transplantation of induced pluripotent stem cells with fibrin glue. Stem Cells Dev 2010;19(11):1757–67.

[79] Bouyer C, Chen P, Güven S, Demirtaş TT, Nieland TJ, Padilla F, et al. A bio-acoustic levitational (BAL) assembly method for engineering of multilayered, 3D brain-like constructs, using human embryonic stem cell derived neuro-progenitors. Adv Mater 2016;28(1):161–7.

[80] Tibbitts D, Rao RR, Shin S, West FD, Stice SL. Uniform adherent neural progenitor populations from rhesus embryonic stem cells. Stem Cells Dev 2006;15(2):200–8.

[81] Clark AR, Carter AB, Hager LE, Price EM. In vivo neural tissue engineering: Cylindrical biocompatible hydrogels that create new neural tracts in the adult mammalian brain. Stem Cells Dev 2016;25(15):1109–18.

[82] Alovskaya A, Alekseeva T, Phillips J, King V, Brown R. Fibronectin, collagen, fibrin-components of extracellular matrix for nerve regeneration. Topics Tissue Eng 2007;3:1–26.

[83] Tate MC, Shear DA, Hoffman SW, Stein DG, Archer DR, LaPlaca MC. Fibronectin promotes survival and migration of primary neural stem cells transplanted into the traumatically injured mouse brain. Cell Transplant 2002;11(3):283–95.

[84] Kearns S, Laywell E, Kukekov V, Steindler D. Extracellular matrix effects on neurosphere cell motility. Exp Neurol 2003;182(1):240–4.

[85] Rappa G, Kunke D, Holter J, Diep D, Meyer J, Baum C, et al. Efficient expansion and gene transduction of mouse neural stem/progenitor cells on recombinant fibronectin. Neuroscience 2004;124(4):823–30.

[86] Xu H, Fan X, Wu X, Tang J, Yang H. Neural precursor cells differentiated from mouse embryonic stem cells relieve symptomatic motor behavior in a rat model of Parkinson's disease. Biochem Biophys Res Commun 2004;326(1):115–22.

[87] Hu J, Deng L, Wang X, Xu XM. Effects of extracellular matrix molecules on the growth properties of oligodendrocyte progenitor cells in vitro. J Neurosci Res 2009;87(13):2854–62.
[88] Jose A, Krishnan LK. Effect of matrix composition on differentiation of nestin-positive neural progenitors from circulation into neurons. J Neural Eng 2010;7(3):036009.
[89] Zychowicz M, Mehn D, Ruiz A, Colpo P, Rossi F, Frontczak-Baniewicz M, et al. Proliferation capacity of cord blood derived neural stem cell line on different micro-scale biofunctional domains. Acta Neurobiol Exp (Wars) 2011;71(1):12–23.
[90] Zeng X, Ma Y, Yf C, Xc Q, Jl W, Ling EA, et al. Autocrine fibronectin from differentiating mesenchymal stem cells induces the neurite elongation in vitro and promotes nerve fiber regeneration in transected spinal cord injury. J Biomed Mater Res A 2016;104(8):1902–11.
[91] Vepari C, Kaplan DL. Silk as a biomaterial. Prog Polym Sci 2007;32(8):991–1007.
[92] Brooks AE. The potential of silk and silk-like proteins as natural mucoadhesive biopolymers for controlled drug delivery. Front Chem 2015;3.
[93] Yang Y, Yuan X, Ding F, Yao D, Gu Y, Liu J, et al. Repair of rat sciatic nerve gap by a silk fibroin-based scaffold added with bone marrow mesenchymal stem cells. Tissue Eng A 2011;17(17–18):2231–44.
[94] Wang J, Ye R, Wei Y, Wang H, Xu X, Zhang F, et al. The effects of electrospun TSF nanofiber diameter and alignment on neuronal differentiation of human embryonic stem cells. J Biomed Mater Res A 2012;100(3):632–45.
[95] Qu J, Wang D, Wang H, Dong Y, Zhang F, Zuo B, et al. Electrospun silk fibroin nanofibers in different diameters support neurite outgrowth and promote astrocyte migration. J Biomed Mater Res A 2013;101(9):2667–78.
[96] Hu A, Zuo B, Zhang F, Lan Q, Zhang H. Electrospun silk fibroin nanofibers promote Schwann cell adhesion, growth and proliferation. Neural Regen Res 2012;7(15):1171.
[97] Lewicka M, Hermanson O, Rising AU. Recombinant spider silk matrices for neural stem cell cultures. Biomaterials 2012;33(31):7712–7.
[98] Manchineella S, Thrivikraman G, Basu B, Govindaraju T. Surface-functionalized silk fibroin films as a platform to guide neuron-like differentiation of human mesenchymal stem cells. ACS Appl Mater Interfaces 2016;8(35):22849–59.
[99] Bai S, Zhang W, Lu Q, Ma Q, Kaplan DL, Zhu H. Silk nanofiber hydrogels with tunable modulus to regulate nerve stem cell fate. J Mater Chem B 2014;2(38):6590–600.
[100] Sun W, Incitti T, Migliaresi C, Quattrone A, Casarosa S, Motta A. Viability and neuronal differentiation of neural stem cells encapsulated in silk fibroin hydrogel functionalized with an IKVAV peptide. J Tissue Eng Regen Med 2015;1–10.
[101] Subia B, Rao RR, Kundu SC. Silk 3D matrices incorporating human neural progenitor cells for neural tissue engineering applications. Polym J 2015;47(12):819–25.
[102] Kubinová Š, Horák D, Vaněček V, Plichta Z, Proks V, Syková E. The use of new surface-modified poly (2-hydroxyethyl methacrylate) hydrogels in tissue engineering: Treatment of the surface with fibronectin subunits versus ac-CGGASIKVAVS-OH, cysteine, and 2-mercaptoethanol modification. J Biomed Mater Res A 2014;102(7):2315–23.
[103] Flanagan LA, Rebaza LM, Derzic S, Schwartz PH, Monuki ES. Regulation of human neural precursor cells by laminin and integrins. J Neurosci Res 2006;83(5):845–56.
[104] Recknor JB, Sakaguchi DS, Mallapragada SK. Directed growth and selective differentiation of neural progenitor cells on micropatterned polymer substrates. Biomaterials 2006;27(22):4098–108.
[105] Nakajima M, Ishimuro T, Kato K, Ko I-K, Hirata I, Arima Y, et al. Combinatorial protein display for the cell-based screening of biomaterials that direct neural stem cell differentiation. Biomaterials 2007;28(6):1048–60.

[106] Poulain FE, Yost HJ. Heparan sulfate proteoglycans: a sugar code for vertebrate development? Development 2015;142(20):3456–67.
[107] Manivasagan P, Oh J. Marine polysaccharide-based nanomaterials as a novel source of nanobiotechnological applications. Int J Biol Macromol 2016;82:315–27.
[108] Bellamkonda R, Ranieri JP, Bouche N, Aebischer P. Hydrogel-based three-dimensional matrix for neural cells. J Biomed Mater Res 1995;29(5):663–71.
[109] Zheng X-S, Yang X-F, Liu W-G, Shen G, Pan D-S, Luo M. A novel method for culturing neural stem cells. In Vitro Cell Develop Biol Anim 2007;43(5–6):155–8.
[110] Birenboim R, Markus A, Goldstein RS. Simple generation of neurons from human embryonic stem cells using agarose multiwell dishes. J Neurosci Methods 2013;214(1):9–14.
[111] Park K, Nam Y, Choi Y. An agarose gel-based neurosphere culture system leads to enrichment of neuronal lineage cells in vitro. In Vitro Cell Dev Biol Anim 2015;51(5):455–62.
[112] Stokols S, Sakamoto J, Breckon C, Holt T, Weiss J, Tuszynski MH. Templated agarose scaffolds support linear axonal regeneration. Tissue Eng 2006;12(10):2777–87.
[113] Aizawa Y, Leipzig N, Zahir T, Shoichet M. The effect of immobilized platelet derived growth factor AA on neural stem/progenitor cell differentiation on cell-adhesive hydrogels. Biomaterials 2008;29(35):4676–83.
[114] Wylie RG, Shoichet MS. Three-dimensional spatial patterning of proteins in hydrogels. Biomacromolecules 2011;12(10):3789–96.
[115] Wong K, Ayuso-Sacido A, Ahyow P, Darling A, Boockvar JA, Wu M. Assessing neural stem cell motility using an agarose gel-based microfluidic device. J Vis Exp. 2008(12):e674-e.
[116] Bidarra SJ, Barrias CC, Granja PL. Injectable alginate hydrogels for cell delivery in tissue engineering. Acta Biomater 2014;10(4):1646–62.
[117] Limoli DH, Jones CJ, Wozniak DJ. Bacterial extracellular polysaccharides in biofilm formation and function. Microbiol Spectr 2015;3(3).
[118] Prang P, Muller R, Eljaouhari A, Heckmann K, Kunz W, Weber T, et al. The promotion of oriented axonal regrowth in the injured spinal cord by alginate-based anisotropic capillary hydrogels. Biomaterials 2006;27(19):3560–9.
[119] Li X, Liu T, Song K, Yao L, Ge D, Bao C, et al. Culture of neural stem cells in calcium alginate beads. Biotechnol Prog 2006;22(6):1683–9.
[120] Ashton RS, Banerjee A, Punyani S, Schaffer DV, Kane RS. Scaffolds based on degradable alginate hydrogels and poly (lactide-co-glycolide) microspheres for stem cell culture. Biomaterials 2007;28(36):5518–25.
[121] Banerjee A, Arha M, Choudhary S, Ashton RS, Bhatia SR, Schaffer DV, et al. The influence of hydrogel modulus on the proliferation and differentiation of encapsulated neural stem cells. Biomaterials 2009;30(27):4695–9.
[122] Matyash M, Despang F, Mandal R, Fiore D, Gelinsky M, Ikonomidou C. Novel soft alginate hydrogel strongly supports neurite growth and protects neurons against oxidative stress. Tissue Eng A 2011;18(1–2):55–66.
[123] Purcell E, Seymour J, Yandamuri S, Kipke D. In vivo evaluation of a neural stem cell-seeded prosthesis. J Neural Eng 2009;6(2):026005.
[124] Purcell EK, Singh A, Kipke DR. Alginate composition effects on a neural stem cell–seeded scaffold. Tissue Eng C Methods 2009;15(4):541–50.
[125] Matos MA, Cicerone MT. Alternating current electric field effects on neural stem cell viability and differentiation. Biotechnol Prog 2010;26(3):664–70.
[126] Li L, Davidovich AE, Schloss JM, Chippada U, Schloss RR, Langrana NA, et al. Neural lineage differentiation of embryonic stem cells within alginate microbeads. Biomaterials 2011;32(20):4489–97.

[127] Bozza A, Coates EE, Incitti T, Ferlin KM, Messina A, Menna E, et al. Neural differentiation of pluripotent cells in 3D alginate-based cultures. Biomaterials 2014;35(16):4636–45.
[128] Salehi H, Karbalaie K, Salamian A, Kiani A, Razavi S, Nasr-Esfahani MH, et al. Differentiation of human ES cell-derived neural progenitors to neuronal cells with regional specific identity by co-culturing of notochord and somite. Stem Cell Res 2012;8(1):120–33.
[129] Khosravizadeh Z, Razavi S, Bahramian H, Kazemi M. The beneficial effect of encapsulated human adipose-derived stem cells in alginate hydrogel on neural differentiation. J Biomed Mater Res B Appl Biomater 2014;102(4):749–55.
[130] Onoe H, Kato-Negishi M, Itou A, Takeuchi S. Differentiation induction of mouse neural stem cells in hydrogel tubular microenvironments with controlled tube dimensions. Adv Healthcare Mater 2016;5(9):1104–11.
[131] Hosseini SM, Sharafkhah A, Koohi-Hosseinabadi O, Semsar-Kazerooni M. Transplantation of neural stem cells cultured in alginate scaffold for spinal cord injury in rats. Asian Spine J 2016;10(4):611–8.
[132] Hunt NC, Hallam D, Karimi A, Mellough CB, Chen J, Steel DH, et al. 3D culture of human pluripotent stem cells in RGD-alginate hydrogel improves retinal tissue development. Acta Biomater 2016;49:329–43.
[133] Ouyang L, Yao R, Zhao Y, Sun W. Effect of bioink properties on printability and cell viability for 3D bioplotting of embryonic stem cells. Biofabrication 2016;8(3):035020.
[134] Fernandes TG, Kwon SJ, Bale SS, Lee MY, Diogo MM, Clark DS, et al. Three-dimensional cell culture microarray for high-throughput studies of stem cell fate. Biotechnol Bioeng 2010;106(1):106–18.
[135] Meli L, Barbosa HS, Hickey AM, Gasimli L, Nierode G, Diogo MM, et al. Three dimensional cellular microarray platform for human neural stem cell differentiation and toxicology. Stem Cell Res 2014;13(1):36–47.
[136] Alessandri K, Feyeux M, Gurchenkov B, Delgado C, Trushko A, Krause K-H, et al. A 3D printed microfluidic device for production of functionalized hydrogel microcapsules for culture and differentiation of human neuronal stem cells (hNSC). Lab Chip 2016;16(9):1593–604.
[137] Croisier F, Jérôme C. Chitosan-based biomaterials for tissue engineering. Eur Polym J 2013;49(4):780–92.
[138] Ahmadi F, Oveisi Z, Samani SM, Amoozgar Z. Chitosan based hydrogels: characteristics and pharmaceutical applications. Res Pharmaceut Sci 2015;10(1):1.
[139] Muzzarelli RA, El Mehtedi M, Bottegoni C, Gigante A. Physical properties imparted by genipin to chitosan for tissue regeneration with human stem cells. Int J Biol Macromol 2016;93(Pt B):1366–81.
[140] Wang G, Ao Q, Gong K, Wang A, Zheng L, Gong Y, et al. The effect of topology of chitosan biomaterials on the differentiation and proliferation of neural stem cells. Acta Biomater 2010;6(9):3630–9.
[141] Hung C-H, Lin Y-L, Young T-H. The effect of chitosan and PVDF substrates on the behavior of embryonic rat cerebral cortical stem cells. Biomaterials 2006;27(25):4461–9.
[142] Soria JM, Martínez Ramos C, Salmerón Sánchez M, Benavent V, Campillo Fernández A, Gómez Ribelles JL, et al. Survival and differentiation of embryonic neural explants on different biomaterials. J Biomed Mater Res A 2006;79(3):495–502.
[143] Nomura H, Zahir T, Kim H, Katayama Y, Kulbatski I, Morshead CM, et al. Extramedullary chitosan channels promote survival of transplanted neural stem and progenitor cells and create a tissue bridge after complete spinal cord transection. Tissue Eng A 2008;14(5):649–65.

[144] Zahir T, Nomura H, Guo XD, Kim H, Tator C, Morshead C, et al. Bioengineering neural stem/progenitor cell-coated tubes for spinal cord injury repair. Cell Transplant 2008;17(3):245–54.
[145] Bozkurt G, Mothe AJ, Zahir T, Kim H, Shoichet MS, Tator CH. Chitosan channels containing spinal cord-derived stem/progenitor cells for repair of subacute spinal cord injury in the rat. Neurosurgery 2010;67(6):1733–44.
[146] Guo X, Zahir T, Mothe A, Shoichet MS, Morshead CM, Katayama Y, et al. The effect of growth factors and soluble Nogo-66 receptor protein on transplanted neural stem/progenitor survival and axonal regeneration after complete transection of rat spinal cord. Cell Transplant 2012;21(6):1177–97.
[147] Li H, Ham TR, Neill N, Farrag M, Mohrman AE, Koenig AM, et al. A hydrogel bridge incorporating immobilized growth factors and neural stem/progenitor cells to treat spinal cord injury. Adv Healthcare Mater 2016;5(7):802–12.
[148] Yang Z, Zhang A, Duan H, Zhang S, Hao P, Ye K, et al. NT3-chitosan elicits robust endogenous neurogenesis to enable functional recovery after spinal cord injury. Proc Natl Acad Sci 2015;112(43):13354–9.
[149] Li X, Yang Z, Zhang A. The effect of neurotrophin-3/chitosan carriers on the proliferation and differentiation of neural stem cells. Biomaterials 2009;30(28):4978–85.
[150] Yang Z, Duan H, Mo L, Qiao H, Li X. The effect of the dosage of NT-3/chitosan carriers on the proliferation and differentiation of neural stem cells. Biomaterials 2010;31(18):4846–54.
[151] Leipzig ND, Xu C, Zahir T, Shoichet MS. Functional immobilization of interferon-gamma induces neuronal differentiation of neural stem cells. J Biomed Mater Res A 2010;93(2):625–33.
[152] Leipzig ND, Shoichet MS. The effect of substrate stiffness on adult neural stem cell behavior. Biomaterials 2009;30(36):6867–78.
[153] Leipzig ND, Wylie RG, Kim H, Shoichet MS. Differentiation of neural stem cells in three-dimensional growth factor-immobilized chitosan hydrogel scaffolds. Biomaterials 2011;32(1):57–64.
[154] Li H, Wijekoon A, Leipzig ND. 3D differentiation of neural stem cells in macroporous photopolymerizable hydrogel scaffolds. PLoS ONE 2012;7(11):e48824.
[155] Li H, Koenig AM, Sloan P, Leipzig ND. In vivo assessment of guided neural stem cell differentiation in growth factor immobilized chitosan-based hydrogel scaffolds. Biomaterials 2014;35(33):9049–57.
[156] Shi W, Nie D, Jin G, Chen W, Xia L, Wu X, et al. BDNF blended chitosan scaffolds for human umbilical cord MSC transplants in traumatic brain injury therapy. Biomaterials 2012;33(11):3119–26.
[157] Zheng L, Cui H-F. Use of chitosan conduit combined with bone marrow mesenchymal stem cells for promoting peripheral nerve regeneration. J Mater Sci Mater Med 2010;21(5):1713–20.
[158] Hsueh Y-Y, Chang Y-J, Huang T-C, Fan S-C, Wang D-H, Chen J-JJ, et al. Functional recoveries of sciatic nerve regeneration by combining chitosan-coated conduit and neurosphere cells induced from adipose-derived stem cells. Biomaterials 2014;35(7):2234–44.
[159] Yi X, Jin G, Tian M, Mao W, Qin J. Porous chitosan scaffold and ngf promote neuronal differentiation of neural stem cells in vitro. Neuro Endocrinol Lett 2010;32(5):705–10.
[160] Jang SJ, Kim YH, Kim HJ, Shim IK, Jeong JM, Chung J-K, et al. Real-time in vivo monitoring of viable stem cells implanted on biocompatible scaffolds. Eur J Nucl Med Mol Imaging 2008;35(10):1887–98.

[161] de Oliveira Barud HG, da Silva RR, da Silva Barud H, Tercjak A, Gutierrez J, Lustri WR, et al. A multipurpose natural and renewable polymer in medical applications: bacterial cellulose. Carbohydr Polym 2016;153:406–20.
[162] Dumanlı A. Nanocellulose and its composites for biomedical applications. *Curr Med Chem* 2016.
[163] Mertaniemi H, Escobedo-Lucea C, Sanz-Garcia A, Gandía C, Mäkitie A, Partanen J, et al. Human stem cell decorated nanocellulose threads for biomedical applications. Biomaterials 2016;82:208–20.
[164] Gu H, Yue Z, Leong WS, Nugraha B, Tan LP. Control of in vitro neural differentiation of mesenchymal stem cells in 3D macroporous, cellulosic hydrogels. Regen Med 2010;5(2):245–53.
[165] Geisel N, Clasohm J, Shi X, Lamboni L, Yang J, Mattern K, et al. Microstructured multilevel Bacterial cellulose allows the guided growth of neural stem cells. Small 2016;12(39):5407–13.
[166] Stabenfeldt SE, Munglani G, García AJ, LaPlaca MC. Biomimetic microenvironment modulates neural stem cell survival, migration, and differentiation. Tissue Eng A 2010;16(12):3747–58.
[167] Sun G, Mao JJ. Engineering dextran-based scaffolds for drug delivery and tissue repair. Nanomedicine 2012;7(11):1771–84.
[168] Huang G, Chen Y, Li Y, Huang D, Han J, Yang M. Two important polysaccharides as carriers for drug delivery. Mini-Rev Med Chem 2015;15(13):1103–9.
[169] Hussain A, Zia KM, Tabasum S, Noreen A, Ali M, Iqbal R, et al. Blends and composites of exopolysaccharides; properties and applications: a review. Int J Biol Macromol 2017;94:10–27.
[170] Sequeira H, Poulain P, Ba-M'Hamed S, Viltart O. Immunocytochemical detection of Fos protein combined with anterograde tract-tracing using biotinylated dextran. Brain Res Protocol 2000;5(1):49–56.
[171] Jurga M, Dainiak MB, Sarnowska A, Jablonska A, Tripathi A, Plieva FM, et al. The performance of laminin-containing cryogel scaffolds in neural tissue regeneration. Biomaterials 2011;32(13):3423–34.
[172] McKee C, Perez-Cruet M, Chavez F, Chaudhry GR. Simplified three-dimensional culture system for long-term expansion of embryonic stem cells. World J Stem Cells 2015;7(7):1064.
[173] Abaskharoun M, Bellemare M, Lau E, Margolis RU. Expression of hyaluronan and the hyaluronan-binding proteoglycans neurocan, aggrecan, and versican by neural stem cells and neural cells derived from embryonic stem cells. Brain Res 2010;1327:6–15.
[174] Wang T-W, Spector M. Development of hyaluronic acid-based scaffolds for brain tissue engineering. Acta Biomater 2009;5(7):2371–84.
[175] Pan L, Ren Y, Cui F, Xu Q. Viability and differentiation of neural precursors on hyaluronic acid hydrogel scaffold. J Neurosci Res 2009;87(14):3207–20.
[176] Seidlits SK, Khaing ZZ, Petersen RR, Nickels JD, Vanscoy JE, Shear JB, et al. The effects of hyaluronic acid hydrogels with tunable mechanical properties on neural progenitor cell differentiation. Biomaterials 2010;31(14):3930–40.
[177] Liang Y, Walczak P, Bulte JW. The survival of engrafted neural stem cells within hyaluronic acid hydrogels. Biomaterials 2013;34(22):5521–9.
[178] Lam J, Carmichael ST, Lowry WE, Segura T. Hydrogel design of experiments methodology to optimize hydrogel for iPSC-NPC culture. Adv Healthcare Mater 2015;4(4):534–9.

[179] Lam J, Lowry WE, Carmichael ST, Segura T. Delivery of iPS-NPCs to the stroke cavity within a hyaluronic acid matrix promotes the differentiation of transplanted cells. Adv Funct Mater 2014;24(44):7053–62.

[180] Park J, Lim E, Back S, Na H, Park Y, Sun K. Nerve regeneration following spinal cord injury using matrix metalloproteinase-sensitive, hyaluronic acid-based biomimetic hydrogel scaffold containing brain-derived neurotrophic factor. J Biomed Mater Res A 2010;93(3):1091–9.

[181] Duan H, Li X, Wang C, Hao P, Song W, Li M, et al. Functional hyaluronate collagen scaffolds induce NSCs differentiation into functional neurons in repairing the traumatic brain injury. Acta Biomater 2016;45:182–95.

[182] Brännvall K, Bergman K, Wallenquist U, Svahn S, Bowden T, Hilborn J, et al. Enhanced neuronal differentiation in a three-dimensional collagen-hyaluronan matrix. J Neurosci Res 2007;85(10):2138–46.

[183] Zhang H, Wei YT, Tsang KS, Sun CR, Li J, Huang H, et al. Implantation of neural stem cells embedded in hyaluronic acid and collagen composite conduit promotes regeneration in a rabbit facial nerve injury model. J Transl Med 2008;6(1):1.

[184] Her GJ, Wu H-C, Chen M-H, Chen M-Y, Chang S-C, Wang T-W. Control of three-dimensional substrate stiffness to manipulate mesenchymal stem cell fate toward neuronal or glial lineages. Acta Biomater 2013;9(2):5170–80.

[185] Lee Y-B, Polio S, Lee W, Dai G, Menon L, Carroll RS, et al. Bio-printing of collagen and VEGF-releasing fibrin gel scaffolds for neural stem cell culture. Exp Neurol 2010;223(2):645–52.

[186] Hiraoka M, Kato K, Nakaji-Hirabayashi T, Iwata H. Enhanced survival of neural cells embedded in hydrogels composed of collagen and laminin-derived cell adhesive peptide. Bioconjug Chem 2009;20(5):976–83.

[187] Lee SH, Chung YN, Kim YH, Kim YJ, Park JP, Kwon DK, et al. Effects of human neural stem cell transplantation in canine spinal cord hemisection. Neurol Res 2009;31(9):996–1002.

[188] Park SS, Lee YJ, Lee SH, Lee D, Choi K, Kim WH, et al. Functional recovery after spinal cord injury in dogs treated with a combination of Matrigel and neural-induced adipose-derived mesenchymal stem cells. Cytotherapy 2012;14(5):584–97.

[189] Liu J, Zhang Z, Gong A, Cao X, Qian L, Duan L, et al. Neuronal progenitor cells seeded in fibrin gel differentiate into ChAT-positive neuron. In Vitro Cell Develop Biol Anim 2010;46(9):738–45.

[190] Carriel V, Garrido-Gomez J, Hernandez-Cortes P, Garzon I, Garcia-Garcia S, Saez-Moreno JA, et al. Combination of fibrin-agarose hydrogels and adipose-derived mesenchymal stem cells for peripheral nerve regeneration. J Neural Eng 2013;10(2):026022.

[191] Carriel V, Scionti G, Campos F, Roda O, Castro B, Cornelissen M, et al. In vitro characterization of a nanostructured fibrin agarose bio-artificial nerve substitute. J Tissue Eng Regen Med 2015;1–10.

[192] Song K, Yang Y, Li S, Wu M, Wu Y, Lim M, et al. In vitro culture and oxygen consumption of NSCs in size-controlled neurospheres of ca-alginate/gelatin microbead. Mater Sci Eng C 2014;40:197–203.

[193] Li W, Guan T, Zhang X, Wang Z, Wang M, Zhong W, et al. The effect of layer-by-layer assembly coating on the proliferation and differentiation of neural stem cells. ACS Appl Mater Interfaces 2015;7(5):3018–29.

[194] Lu HF, Lim S-X, Leong MF, Narayanan K, Toh RP, Gao S, et al. Efficient neuronal differentiation and maturation of human pluripotent stem cells encapsulated in 3D microfibrous scaffolds. Biomaterials 2012;33(36):9179–87.

[195] Gu Q, Tomaskovic-Crook E, Lozano R, Chen Y, Kapsa RM, Zhou Q, et al. Functional 3D neural mini-tissues from printed gel-based Bioink and human neural stem cells. Adv Healthcare Mater 2016;5(12):1428.
[196] Ren Y-J, Zhou Z-Y, Liu B-F, Q-Y X, Cui F-Z. Preparation and characterization of fibroin/hyaluronic acid composite scaffold. Int J Biol Macromol 2009;44(4):372–8.
[197] Shoichet MS, Tator CH, Poon P, Kang C, Baumann MD. Intrathecal drug delivery strategy is safe and efficacious for localized delivery to the spinal cord. Prog Brain Res 2007;161:385–92.
[198] Kang CE, Poon PC, Tator CH, Shoichet MS. A new paradigm for local and sustained release of therapeutic molecules to the injured spinal cord for neuroprotection and tissue repair. Tissue Eng A 2008;15(3):595–604.
[199] Baumann MD, Kang CE, Stanwick JC, Wang Y, Kim H, Lapitsky Y, et al. An injectable drug delivery platform for sustained combination therapy. J Control Release 2009;138(3):205–13.
[200] Wang Y, Lapitsky Y, Kang CE, Shoichet MS. Accelerated release of a sparingly soluble drug from an injectable hyaluronan–methylcellulose hydrogel. J Control Release 2009;140(3):218–23.
[201] Mothe AJ, Tam RY, Zahir T, Tator CH, Shoichet MS. Repair of the injured spinal cord by transplantation of neural stem cells in a hyaluronan-based hydrogel. Biomaterials 2013;34(15):3775–83.
[202] Ballios BG, Cooke MJ, van der Kooy D, Shoichet MS. A hydrogel-based stem cell delivery system to treat retinal degenerative diseases. Biomaterials 2010;31(9):2555–64.
[203] Caicco MJ, Cooke MJ, Wang Y, Tuladhar A, Morshead CM, Shoichet MS. A hydrogel composite system for sustained epi-cortical delivery of Cyclosporin a to the brain for treatment of stroke. J Control Release 2013;166(3):197–202.
[204] Wang Y, Cooke MJ, Sachewsky N, Morshead CM, Shoichet MS. Bioengineered sequential growth factor delivery stimulates brain tissue regeneration after stroke. J Control Release 2013;172(1):1–11.
[205] Wang H, Lin XF, Wang LR, Lin YQ, Wang JT, Liu WY, et al. Decellularization technology in CNS tissue repair. Expert Rev Neurother 2015;15(5):493–500.
[206] Wang Q, Zhang C, Zhang L, Guo W, Feng G, Zhou S, et al. The preparation and comparison of decellularized nerve scaffold of tissue engineering. J Biomed Mater Res A 2014;102(12):4301–8.
[207] Szynkaruk M, Kemp SW, Wood MD, Gordon T, Borschel GH. Experimental and clinical evidence for use of decellularized nerve allografts in peripheral nerve gap reconstruction. Tissue Eng B Rev 2012;19(1):83–96.
[208] Nagao RJ, Lundy S, Khaing ZZ, Schmidt CE. Functional characterization of optimized acellular peripheral nerve graft in a rat sciatic nerve injury model. Neurol Res 2013;33(6):600–8.
[209] Hudson TW, Zawko S, Deister C, Lundy S, Hu CY, Lee K, et al. Optimized acellular nerve graft is immunologically tolerated and supports regeneration. Tissue Eng 2004;10(11–12):1641–51.
[210] Shanti RM, Ziccardi VB. Use of decellularized nerve allograft for inferior alveolar nerve reconstruction: A case report. J Oral Maxillofac Surg 2011;69(2):550–3.
[211] Zhao Z, Wang Y, Peng J, Ren Z, Zhang L, Guo Q, et al. Improvement in nerve regeneration through a decellularized nerve graft by supplementation with bone marrow stromal cells in fibrin. Cell Transplant 2014;23(1):97–110.

[212] Liu F, Lin H, Zhang C. Construction of tissue-engineered nerve conduits seeded with neurons derived from hair-follicle neural crest stem cells. Methods Mol Biol 2016;1453:33–8.

[213] Liu J, Chen J, Liu B, Yang C, Xie D, Zheng X, et al. Acellular spinal cord scaffold seeded with mesenchymal stem cells promotes long-distance axon regeneration and functional recovery in spinal cord injured rats. J Neurol Sci 2013;325(1):127–36.

[214] Karumbaiah L, Enam SF, Brown AC, Saxena T, Betancur MI, Barker TH, et al. Chondroitin sulfate glycosaminoglycan hydrogels create endogenous niches for neural stem cells. Bioconjug Chem 2015;26(12):2336–49.

CHAPTER

Synthetic biomaterials for engineering neural tissue from stem cells

6

1 INTRODUCTION

Naturally derived biomaterials have major limitations in terms of reproducibility and sourcing as detailed in Chapter 5. In contrast, a diverse number of synthetic polymers can be used to generate 3D scaffolds for applications in tissue engineering [1]. These synthetic polymers are produced using standard chemical reactions and thus being able to generate scaffolds from them that provide an alternative way to construct a supportive microenvironment for influencing stem cell behavior, while avoiding the sourcing and reproducibility issues of natural polymers. Synthetic materials are commonly used for biomedical applications with polymers such as poly (ethylene glycol) (PEG) being used as a coating for medical devices to limit the immune response by the body and poly (caprolactone) (PCL) from which sutures are often made. Such polymers often vary in their ability to be processed, enabling the fabrication of scaffolds with a wider variety of structures, including the ability to generate novel topographies such as nanofibers and microfibers [2]. Scaffolds fabricated from synthetic polymers can also be designed for applications in drug delivery [3]. Also, synthetic polymers are not limited to molecules found in nature and thus offer a broader range of mechanical properties as well.

Many synthetic polymers, such as PEG, resist cell adhesion, which may sound like a disadvantage when engineering neural tissue. However, it enables researchers to selectively functionalize scaffolds created from such polymers with sites that promote the binding of cells in selected regions, expanding the possibilities for engineering tissue. For example, these scaffolds made from synthetic materials can be manipulated to present bioactive molecules that enable cell adhesion by using chemistry to attach components derived from natural biomaterials or synthetic peptide sequences that mimic such materials [4]. Accordingly, this chapter is structured by including such studies that combine natural and synthetic polymers under the relevant synthetic biomaterial heading when they are discussed in Section 2. Fig. 1 summarizes these aforementioned properties of synthetic scaffolds that can be manipulated to influence stem cell behavior, which is taken from a recent review by the Burdick group [5].

This chapter begins by reviewing the most commonly used synthetic polymers for engineering neural tissue from stem cells. These polymers include poly (ethylene glyc*ol*), poly (ethylene-co-vinylacetate), poly (glycolic acid)/poly (lactic acid)/poly (lactic-*co*-glycolic acid), poly (caprolactone), and poly (2-hydroxyethyl

Engineering Neural Tissue from Stem Cells. http://dx.doi.org/10.1016/B978-0-12-811385-1.00006-6

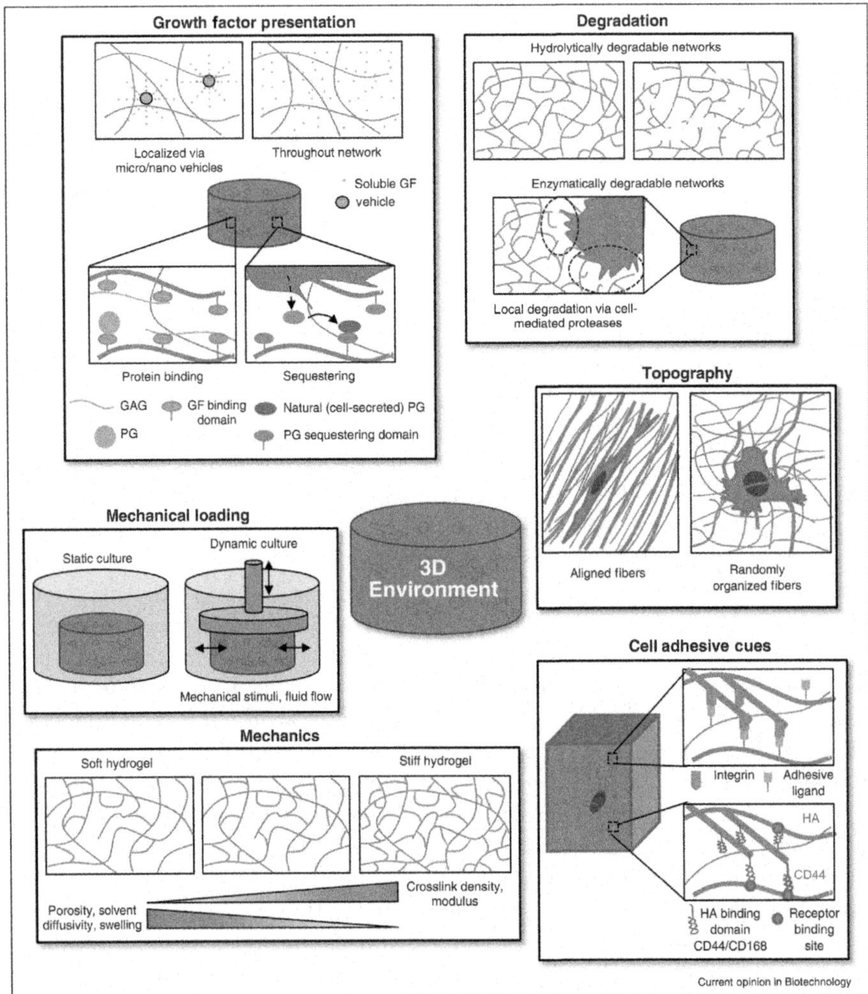

FIG. 1

Stem cells receive cues from numerous properties and features of their culture environment, which provide instruction on outcomes such as proliferation and differentiation. Within synthetic materials, these cues may include material degradation, topography, adhesion, growth factor presentation and mechanics, as well as the application of external loading.

Reprinted with permission from Guvendiren M, Burdick JA. Engineering synthetic hydrogel microenvironments to instruct stem cells. Curr Opin Biotechnol 2013;24(5):841–6.

methacrylate). This chapter will also discuss the use of scaffolds made from synthetic peptides as well. For our purposes, peptides generated using synthesis are considered to be synthetic biomaterials as their amino acid sequence can manipulated during this process to generate sequences that are not found in nature. These peptide-based scaffolds also can be used for neural tissue engineering applications [6]. It then discusses unique combinations of synthetic and natural biomaterials that can be used for these applications as well before providing areas for future research. Finally, the ability to design polymers that response to different stimuli is possible and these responsive materials provide unique opportunities when engineering tissues as detailed in Section 3[7,8]. Suggestions for future work are also given in the final section.

2 TYPES OF SYNTHETIC SCAFFOLDS

A variety of synthetic polymers is already approved by the Food and Drug Administration for use in biomedical devices and for applications in drug delivery as mentioned in the Introduction section. While the Food and Drug Administration technically approves each device or medical product based on its own merits, all biomaterials must conform a set of known international standards with regard to immunogenicity, toxicity, and other properties before they can be tested in humans. Thus, these already FDA-approved synthetic biomaterials provide an excellent starting point for engineering tissues as they have already met these standards, which indicate they do not cause significant cell death or other issues upon in vivo implantation in humans. This chapter begins by discussing one of the most commonly used synthetic biomaterial—PEG—before moving onto other synthetic polymers that have been applied for neural tissue engineering using stem cells.

Electrospinning is a popular fabrication technique used to generate 3D scaffolds from synthetic polymers for a variety of applications, including tissue engineering [9]. This process produces fibers from a highly charged polymer solution. Two variations of electrospinning exist—solution and melt. Solution electrospinning is the more commonly used process. It involves dissolving the desired polymer for scaffold fabrication into a solvent to generate a solution. A feeder syringe then delivers this solution onto a charged plate through spinning, as the difference in charge attracts the polymer—drawing out nanofibers in the process. Solid fibers form on the collector plate. Bioactive components can be added to the polymer solution to enhance the properties of the scaffold produced [10]. The shape of the manner of nanofiber collection can be varied, which can result in a variety of topographies as seen in Fig. 2 [11]. Melt electrospinning, as its name implies, requires the polymer to be heated up until it melts, forming a liquid that can then be spun. These polymer melts are charged and then the resulting solution can be used to write directly 3D scaffolds for tissue engineering applications [12–15]. Melt electrospinning provides more control over the structures that can be fabricated. The fibers produced by melt electrospinning tend to be on the micro range compared to the nanofibers typically produced using solution electrospinning. Both levels of structures are present in native tissue and there are benefits to being able to control the topography in both length scales. However, the use of heat to melt the polymer makes it difficult to incorporate bioactive proteins into such scaffolds, as they would be denaturing during this process. However, some drug releasing microfibers have been generated using melt electrospinning [16–18]. This chapter will discuss studies involving these electrospun scaffolds under the appropriate polymer heading.

FIG. 2

Different types of topography generated using the process of electrospinning.

Reprinted with permission from Tian L, Prabhakaran MP, Ramakrishna S. Strategies for regeneration of components of nervous system: scaffolds, cells and biomolecules. Reg Biomater 2015:rbu017.

Synthetic polymers are also commonly used to fabricate drug-releasing microspheres. These microspheres can be formulated specifically for applications in neural tissue engineering [3]. For example, they can continuously release bioactive growth factors, antibodies, or small molecules over a certain period of time. These factors can either promote regeneration of damaged nervous system or direct stem cells to survive or differentiate into mature neural phenotypes as detailed in Chapter 4. Such microspheres can be incorporated into different types of biomaterial scaffolds to enhance their bioactive properties. Overall, synthetic biomaterials enable the production of versatile structures in terms of both topography and drug delivery. This chapter will now discuss specific applications of such materials for neural tissue engineering applications.

2.1 POLY (ETHYLENE GLYCOL)/POLY (ETHYLENE OXIDE)

Poly (ethylene glycol) (PEG) is one of the most commonly used synthetic biomaterials due to its numerous desirable properties. It is sometimes referred to as poly (ethylene oxide) (PEO) depending on its molecular weight, as PEO usually refers to high-molecular weight version. Its polymer chain consists of monomers of ethylene oxide with the properties of PEG depending on the length of the polymer chain synthesized. PEG can also be produced in a form that possesses multiple arms as common versions have either four or eight arms that branch out from a central site [19]. PEG generally exists as a solution that can then be chemically cross-linked into hydrogel scaffolds for applications in neural tissue engineering [20]. It also demonstrates excellent biocompatibility [21]. PEG also can be readily modified using a variety of chemistries to improve its properties, including making it bioactive [22]. It can be formulated so that to polymerize in situ, which is an attractive feature for in vivo delivery of such scaffolds [23,24]. As mentioned in the Introduction section, PEG resists protein adsorption, hence its application as a coating for medical devices as it prevents proteins that would trigger the immune response from the body from binding to the surface of the device. This property also means that engineering tissues often use chemically modified PEG that contains natural biomaterial components or peptide sequences for supporting cell adhesion and growth. While not all the studies discussed in this section directly use stem cells, these studies highlight the different ways PEG scaffolds can be manipulated to improve their properties for applications in neural tissue engineering. Several of these examples represent bioactive scaffolds that could be adapted for applications using stem cells.

In 1994, Dr. Mark Saltzman and colleagues functionalized PEG with two different peptide domains designed to promote cell adhesion (the amino acid sequences: RGD and YISGR) and then seeded neural cancer cells onto these hydrogels, which were then added to a rotating bioreactor for culture [25]. These scaffolds promoted the aggregation of the neural cells compared to control cultures. In a related work on neural tissue, the Langer group conducted a study led by Dr. Jason Burdick that demonstrated PEG hydrogels combined with the delivery of ciliary neurotrophic factor supported the culture of retinal explants [26]. The Lavik research group has also made significant contributions to the body of knowledge about using PEG hydrogels for applications in neural stem cell culture. In 2007, they developed a photopolymerizable PEG/poly (L lactic)(PLL) acid scaffold that supported neural stem cell differentiation into mature neurons when epidermal growth factor was present in the cell culture media [27]. They further expanded on this work by generating a tunable library of synthetic PEG constructs that modulated neural stem cell behavior mainly through changes in the mechanical properties of these scaffolds [28]. They also created a microporous version of these PEG-PLL scaffolds that supported neural stem cell growth and differentiation while enabling the formation of the vasculature inside the same scaffold simultaneously [29].

Another group explored how modulating the amount of PLL used in a PEG scaffold affected the behavior of mouse neural stem cells seeded inside [30]. They found that 2% PLL content inside of PEG scaffolds promoted neural stem cell proliferation and differentiation. Work from Dr. Jennifer West's group showed that immortalized neural stem cells could be successfully encapsulated inside of PEG microspheres using photopolymerization [31]. These cells maintained a high level of viability post encapsulation and could be released from these scaffolds using enzymes, suggesting this technique could be used for neural stem cell expansion and differentiation. Finally, recent work from the University of Wisconsin used PEG hydrogels in combination with neurons derived from human-induced pluripotent stem cells as a way to screen compounds for neurotoxicity [32], demonstrating how such scaffolds can support culture of pluripotent stem cells as well.

A rational approach for designing such scaffolds screened PEG arrays functionalized with different cues to determine their effect on neural stem cell behavior and then used the resulting data to produce an optimized PEG scaffold for neural tissue engineering applications [33]. Among the cues identified in this screening process were laminin, a fibronectin domain, and the jagged ligand that activate Notch signaling, suggesting that their incorporation would be beneficial for producing an optimal bioactive PEG hydrogel formulation for neural stem cell culture. Other groups have used PEG in combination with dextran to create microengineered niches for mouse embryonic stem cells using a 2 phase aqueous system that includes coculture with Pa6 stromal cells [34]. Such a system could also be used to screen cues to determine their influence on stem cell differentiation. Another group developed a PEG hydrogel formulation that gelled upon injection in the body [35]. These scaffolds were modified with either gelatin or hyaluronic acid to improve their cell binding capabilities. They used this novel scaffold to deliver oligodendrocyte progenitors into the injured spinal cord where the gel increased cell survival and the transplanted cells then matured into oligodendrocytes. Follow-up work used this hydrogel formulation for supporting neural stem cell growth and differentiation in vitro as well [36].

Another interesting study showed that PEG hydrogels containing varying concentrations of hyaluronic acid had markedly different effects on neural stem cells depending on whether the stem cells were isolated from a fetus or an adult [37]. This work suggests that different hydrogel formulations should be used for these different types of stem cells. In other related work, the Mahoney group demonstrated that PEG hydrogels with faster degradation rates enhanced the proliferation of multipotent neural precursor cells and neurons [38]. A follow-up study showed that generating PEG scaffold from lower macromer weights increased metabolic activity while decreasing apoptosis for neural cells [39]. The first author on this study, Dr. Kyle Lampe, now works as a professor at the University of Virginia. Further work from the Mahoney group showed that the behavior of neural stem cells seeded inside of 3D PEG scaffolds can be controlled through treatment with soluble factors such as basic fibroblast growth factor and epidermal growth factor [40]. They also found that the 3D hydrogels were more supportive of neural precursor cells than traditional 2D monolayer culture.

A pair of studies using a combination of PEG and methyl alginate and a combination of PEG and collagen demonstrated how modulating the mechanical properties can influence neural cancer cell behavior [41,42]. The first study showed that stiffness increased cell proliferation while the second study indicated that increasing the concentration of collagen present in a PEG gel enhanced neurite extension from cells. Other work has further characterized how modulating the concentrations of fibronectin and collagen incorporated into PEG scaffolds and characterized how they affected hydrogel stiffness and cell behavior as both these properties influenced neurite extension [42]. These observations can be translated for optimizing the design of PEG scaffolds for stem cell applications.

An interesting approach to functionalizing different natural proteins requires chemically conjugating them to PEG in a process called PEGylation, which results in a uniquely modified protein [43]. Both PEGylated fibrin and PEGylated collagen can be formulated and hydrogels made from these unique materials support neural cell culture when neurotrophins were present in the cell culture media [44]. Another approach used fibrin to deliver primary neural cells into PEG hydrogels to make these scaffolds suitable for neural cell culture [38]. Once the fibrin was degraded, it left behind a porous network where the neural cells migrated and extended neurites throughout the remaining scaffold. PEGylation can also be used to functionalize regeneration inducing drugs. For example, a PEGylated version of an antibody against the protein LINGO caused increased myelination in the injured spinal cord with greater efficacy than the unmodified version [45]. Thus, in addition to being used as a hydrogel scaffold—PEG can also be used to produce more potent versions of both naturally derived biomaterials and protein-based therapeutics.

In terms of drug delivery from PEG scaffolds, different group have demonstrated that such scaffolds can deliver several relevant growth factors. For example, the Willets group used PEG scaffolds to deliver bioactive nerve growth factor (NGF) over a 20-day time course [46]. Another study from the Elbert group created unique scaffolds from heparinized PEG microspheres that delivered glial-derived neurotrophic factor for applications in engineering neural tissue [47]. They could also manipulate such scaffolds to generate gradients of the released protein. Follow-up work evaluated nerve guidance conduits fabricated from such a system in an in vivo rat sciatic nerve injury model where they promoted robust axonal regeneration [48].

Another sophisticated fabrication technique used photolithography to generate complex 3D scaffolds for neural cell culture that consisted of PEG regions that did not permit cell adhesion along with regions of Puramatrix (a synthetic peptide scaffold discussed in more depth in Section 2.6) or agarose [49]. While this work seeded these uniquely structured scaffolds with dorsal root ganglion, these scaffolds could also be used in combination with stem cells. In similar work, a complex nerve guidance conduits containing single or multiple lumens were fabricated from PEG using stereolithography, a 3D printing process that converts a liquid ink into a solid object [50]. The single lumen conduits were more effective at promoting nerve regeneration in a 10-mm gap in rats, suggesting the importance of scaffold structure in promoting regeneration.

Finally, recent studies have been combining defined PEG scaffolds with pluripotent stem cells as ways to engineer organoids from embryonic stem cells [51]. Organoids are miniature tissues that replicate the structures found in native tissues in the human body. In this study, mouse embryonic stem cells embedded in PEG hydrogels were treated with retinoic acid and these aggregates formed neuroepithelial cysts as seen in Fig. 3. Treatment with retinoic acid generated local expression of the protein sonic hedgehog, causing these cysts to polarize in a manner similar to how the neural tube is patterned during development. This book will address the topic of organoids for neural tissue applications in more depth in Chapter 8.

The work detailed in this section illustrates how versatile PEG is as a biomaterial. It can be manipulated to contain sites for cell adhesion and enhance neurite extension and its mechanical properties can be altered to match those of native neural tissue.

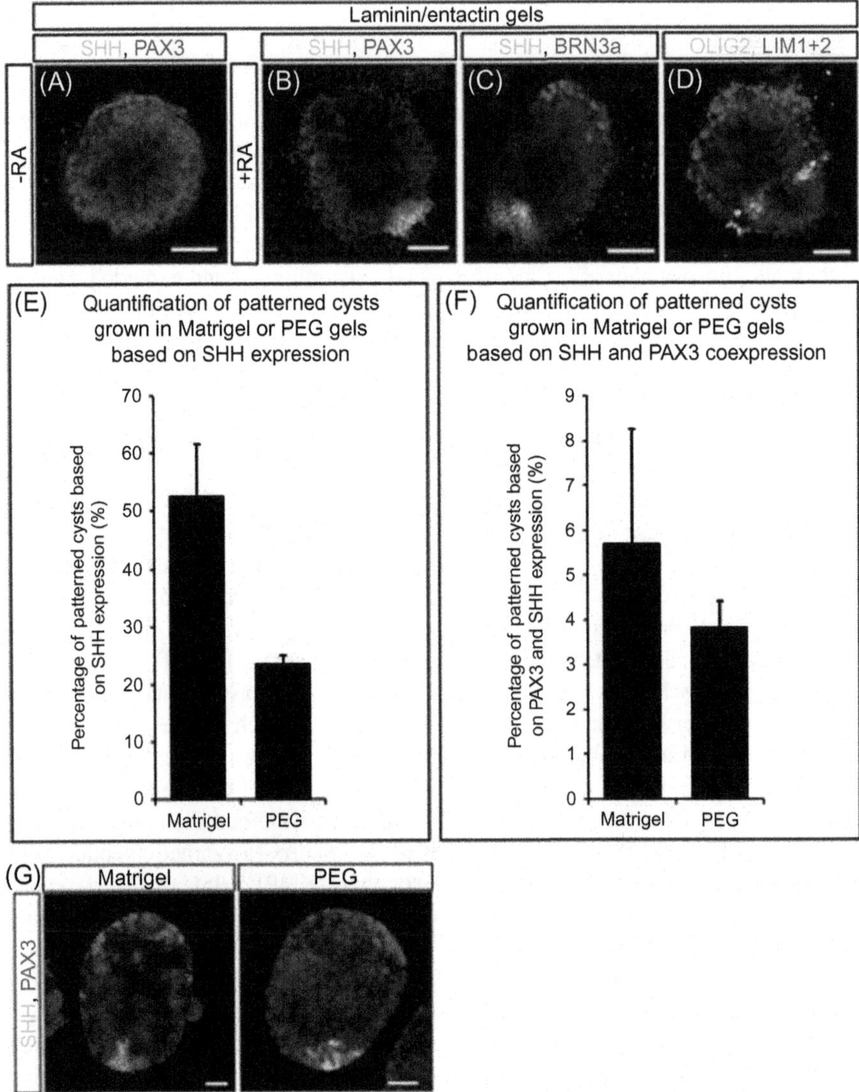

FIG. 3

Defined 3D matrices also permit RA-induced self-patterning of neuroepithelial cysts. (A–D) Cysts were grown in high concentration laminin/entactin gels. Untreated control cysts were uniformly PAX3+ (*red in the electronic version, gray in the print version*, A) and patterned similar to cysts grown in Matrigel upon a RA pulse on day 2 as evidenced by immunofluorescence staining for SHH *(green)* and PAX3 (*red*, B), SHH *(green)* and BRN3a (*red in the electronic version, gray in the print version*, C) as well as OLIG2 *(green in the electronic version, gray in the print version)* and LIM1+2 (*red in the electronic version, gray in the print version*, D). (E) Even in pure PEG gels, cysts grew and patterned after RA administration based on focal SHH expression, although to a lesser extent than in Matrigel as quantified. (F and G) Complete DV patterned cysts based on coexpression of PAX3 and SHH were also observed. Data are represented as mean ± SD for Matrigel ($n=3$ independent experiments) and mean ± variance for PEG gels ($n=2$ independent experiments) with 100 cysts counted per experiment. Nuclei were counterstained with Hoechst. Scale bars represent 50 μm (A–D) and 20 μm (G).

Reprinted with permission from Meinhardt A, Eberle D, Tazaki A, Ranga A, Niesche M, Wilsch-Bräuninger M, et al. 3D reconstitution of the patterned neural tube from embryonic stem cells. Stem Cell Rep 2014;3(6):987–99.

PEG scaffolds can deliver growth factors in a controlled fashion over an extended time course. Finally, it can produce scaffolds possessing a wide variety of shapes depending on the fabrication technique used. The main challenge is how to combine these different properties into a single scaffold formulation that successful presents both chemical and mechanical cues for supporting stem cell differentiation into functional neural tissue.

2.2 POLY (GLYCOLIC ACID)/POLY (LACTIC ACID)/POLY (LACTIC-*CO*-GLYCOLIC ACID)

This section will discuss two polymers, (poly (glycolic acid) (PGA) and poly (lactic acid), and their blends for applications in neural tissue using stem cells. PGA consists of glycolic acid monomers, while PLA consists of lactic acid monomers. They are often blended together to make poly (lactic-*co*-glycolic acid), which is commonly used for biomedical applications due to its inherent biocompatibility and biodegradability [52]. These blends are often characterized by their molar ratios of lactic acid to glycolic acid. These ratios influence the properties of the resulting polymer, including its glass transition temperature where it converts from a hard polymer into a rubber material. PLGA, like PEG, can be manipulated into a variety of structures. For example, PLGA can be easily fabricated into nanofibers and microspheres [53,54]. One potential concern when using PLGA scaffolds is that water breaks down the ester bonds in the polymer backbone, which results in the byproducts that are toxic to cells.

Initial work evaluating the use of PLA, PGA, and PLGA scaffolds for use with neural stem cell culture was performed in the Langer group and was led by Dr. Lavik. In 2002, they reported that meshes made from these polymers could be used as substrates for culturing neural stem cells and this report included detailed protocols for the entire process [55]. Follow-up work confirmed that using PLGA scaffolds in combination with retinoic acid promoted neuronal differentiation of human embryonic stem cells [56]. Treating human embryonic stem cells seeded inside of PLGA scaffolds with neurotrophins, such as nerve growth factor (NGF) and neurotrophin-3 (NT-3), resulted in neural tissue that also contained vasculature [57]. In other work, PLGA scaffolds seeded with bone marrow stromal cells could be used to repair peripheral nerve injuries in rats [58]. These stromal cells differentiated into Schwann-like cells that contributed to neural regeneration. Another study compared various synthetic polymers for their ability to culture neural progenitor cells derived from the hippocampus, and it was found the PLGA was the best in terms of maintaining cell viability [59]. Another interesting approach to engineering neural tissue involved seeding PLGA scaffolds with mesenchymal stem cells modified to express NGF [60]. The secreted NGF was biologically active. Other research confirmed that neural stem cells could form neural networks inside of 3D PLGA scaffolds as the seeded cells expressed several synaptic markers [61]. Several studies have demonstrated the suitability of PLGA for delivering a variety of cell types into the injured spinal cord [62–65]. These cell types include neural stem cells, endothelial progenitor cells, and Schwann cells, demonstrating the versatility of these scaffolds. PLGA scaffolds have also been used to deliver successfully β-mercaptoethanol and the protein sonic hedgehog in a controlled fashion [66,67].

Micropatterning can also be used to fabricate 3D PLGA structures, enabling generation of more complex tissue architecture when engineering neural tissue from stem cells [68]. For example, PLGA channels can be formed, which promote stem cell differentiation [69]. The effects of such nanotopographical features can also be combined with controlled drug release to synergistically direct the organization and differentiation of human neural stem cells [70]. Other unique topography can be generated in PLGA scaffolds by using salt leaching to create pores throughout the scaffold [71]. These unique scaffolds supported the differentiation of mouse-induced pluripotent stem cells into retinal cells.

As mentioned earlier, fabricating drug-releasing microspheres from PLGA can be an effective strategy for promoting neural regeneration and directing stem cells to differentiate into neural phenotypes. For example, PLGA microspheres can deliver ciliary neurotrophic factor in a controlled fashion to neural progenitors, making them differentiate into astrocytes [72]. Other work used PLGA microspheres to deliver retinoic acid to differentiate stem cells into neurons [73,74]. Such microspheres can also be used as scaffolds to support neural stem cell survival as demonstrated in an in vivo animal model of stroke [75,76]. They can also be incorporated in natural biomaterial scaffolds to increase their bioactivity. For example, hyaluronic acid scaffolds were formulated to contain PLGA microspheres that released both brain-derived neurotrophic factor and vascular endothelial growth factor [77]. The release of these growth factors promoted the survival and proliferation of the neural stem cells seeded inside of the scaffold. The Shoichet group combined PLGA microspheres that released dibutyryl cyclic-AMP with chitosan scaffolds to induce neural stem cells to mature into neurons [78]. These scaffolds were evaluated in a preclinical spinal cord injury model where they promoted high levels of cell survival and differentiation post transplantation. Neurotrophin-3 releasing PLGA microspheres were incorporated into nerve guidance conduits along with neural stem cells and Schwann cells [79]. The stem cells matured into neurons that were then myelinated by the Schwann cells. Other work demonstrated that PLGA microspheres could deliver bioactive glial-derived neurotrophic factor (GDNF) to differentiate neural progenitor cells into mature neurons [80]. Overall, PLGA can be applied in several ways for engineering neural tissue from stem cells, including delivery of biomolecules and present novel topographies.

2.3 POLY (CAPROLACTONE)

Poly (caprolactone) (PCL) possesses many desirable qualities as a biomaterial [81]. It is a low-cost polymer and biocompatible as it is commonly used to fabricate surgical sutures. It has a long degradation rate, meaning that it could potentially deliver cues for differentiating stem cells over extended time periods of months. The PCL polymer is produced by opening the ring of the molecule ε-caprolactone, which can be performed using a variety of catalysts to drive this reaction [82]. One of the desirable properties of PCL is its low melting temperature as it is easy to manipulate and process into different types of structures. PCL has become a very popular material to use when electrospinning fibrous scaffolds for applications in tissue engineering [83]. Accordingly, most of the studies discussed in this section involve seeding different types of stem cells onto electrospun scaffolds for applications in neural tissue engineering.

Electrospinning PCL into nanofiber scaffolds enables the generation of topography similar to that found in the extracellular matrix. This process also allows growth factors and small molecules to be incorporated into these scaffolds during fabrication, adding further functionalization and enhancing their properties. A significant amount of work has investigated how such scaffolds can be used to influence the behavior of stem cells, including their differentiation into neural lineages [9]. In 2008, Dr. David Nisbet and his colleagues in Dr. Shoichet's lab demonstrated how randomly aligned PCL nanofibers treated with ethylenediamine supported the adhesion of neural stem cells [84]. These cells differentiated into mainly oligodendrocytes in the presence of serum, indicating how such substrate can influence differentiation. During my graduate studies, I collaborated with the Xia lab to demonstrate how aligned PCL scaffolds produced using the process of solution electrospinning enhanced the differentiation of mouse embryonic stem cell-derived neural progenitors into neurons while reducing the fraction of cells that differentiated into astrocytes [85]. The neurites extended by the neurons also would migrate along the aligned fibers, providing a method for organizing these cells into tissue-like structures.

My research group at the University of Victoria has also made significant contributions in using PCL electrospun scaffolds for promoting the differentiation of pluripotent stem cells into neural tissue. Our electrospinning work was performed in collaboration with my colleague Dr. Martin Jun, who is now at Purdue University. First, we followed up on this work by showing that we could fabricate aligned scaffolds that would release retinoic acid [86]. We showed that mouse-induced pluripotent stem cell-derived embryoid bodies could adhere and differentiate into neurons on such scaffolds and they also extended neurites along the aligned fibers. We performed similar work to generate aligned PCL nanofiber scaffolds that released bioactive glial-derived neurotrophic factor slowly over 30 days [87]. At the end of 30 days, 7%–10% of the total protein loaded into the scaffolds had been released and the release rate varied with the topography of the scaffolds (random versus aligned). These scaffolds supported the differentiation of neural progenitors derived from human-induced pluripotent stem cells into neurons. We also have looked at the use of PCL microfibers produced using melt electrospinning as substrates for the culture of pluripotent stem cells. We showed that mouse embryonic stem cells could differentiate into neurons on such microfiber scaffolds [88]. Our lab also generated novel scaffolds by combining melt electrospun microfiber scaffolds with drug releasing nanofibers produced using solution electrospinning [89]. This work showed that biaxially aligned microfiber scaffolds enhanced the neurite extension of neural progenitors derived from human-induced pluripotent stem cells. It also showed that these novel combination scaffolds supported stem cell culture and differentiation into neurons. A 2011 study examined the role of fiber size and orientation on the differentiation of human embryonic stem cells into neural phenotypes [90]. Coating such scaffolds with laminin promoted cell adhesion and they found aligned fibers promoted neuronal differentiation, consistent with work from my lab and other groups. Other groups showed how laminin-coated PCL scaffolds could mimic different ECM topographies such as the vasculature and the radial glia, which in turn caused different behaviors from neural stem cells depending on the region of ECM mimicked [91].

Tethering growth factors to the surface of electrospun scaffolds serves as an interesting approach for enhancing their bioactive properties. For example, aligned electrospun PCL scaffolds can be modified so that brain-derived neurotrophic factor tethered to their surface [92,93]. This combination of biophysical and biochemical cues resulted in enhanced differentiation of neural stem cells into neurons and oligodendrocytes. Similar work was performed using mesenchymal stem cells seeded on aligned, retinoic acid releasing PCL scaffolds, which induced these cells to differentiate into neuron-like cells [94]. Other work has tethered glial-derived neurotrophic factor on the surface of such nanofiber scaffolds—demonstrating another effective example of this drug delivery strategy [95]. Such nanofibers can also deliver NGF as well [96]. In this study, these molecules (NGF in combination with retinoic acid) were used to drive the differentiation of mesenchymal stem cells into neural-like cells. PCL nanofibers can also be coated with polypyrrole (an electrically conductive polymer), which enabled more adhesion by neural stem cells while promoting their expansion and differentiation [97]. Collagen is another popular option for increasing the number of adhesion sites for cells on PCL nanofiber scaffolds [98]. Such scaffolds support the culture and differentiation of rat mesenchymal stem cells into neuronal-like phenotypes. Another group added gelatin to PCL nanofibers to generate substrates that differentiated oligodendrocyte progenitor cells into mature oligodendrocytes [99]. Other means of improving cell adhesion include functionalizing PCL with peptides that mimic the protein vimentin [100]. These scaffolds promoted the differentiation of retinal progenitor cells into mature photoreceptors under the appropriate culture conditions. Finally, these PCL nanofibers can be treated with plasma to improve the ability of mesenchymal stem cells to bind to their surface [101], which is another option for improving cytocompatibility.

An interesting study from the Chew group used PCL scaffolds to deliver silencing RNA (siRNA) that knocked down the expression of RE-1 silencing transcription factor in mouse neural progenitor cells, which resulted in enhanced neuronal differentiation [102]. This work demonstrates how PCL scaffolds can serve to manipulate stem cell behavior at the genetic level. Such effects were also combined with nanotopography for differentiating oligodendrocyte progenitor cells [103]. Another novel approach to engineering neural tissue used PCL scaffolds to deliver neural stem cells along with NT-3 and chondroitinases into the injured spinal cords of rats [104]. This combination strategy resulted in functional recovery post injury compared to control animals. Other work used novel PCL nanotubes to deliver adipose-derived stem cells into a peripheral nerve injury model where they were able to promote regeneration [105].

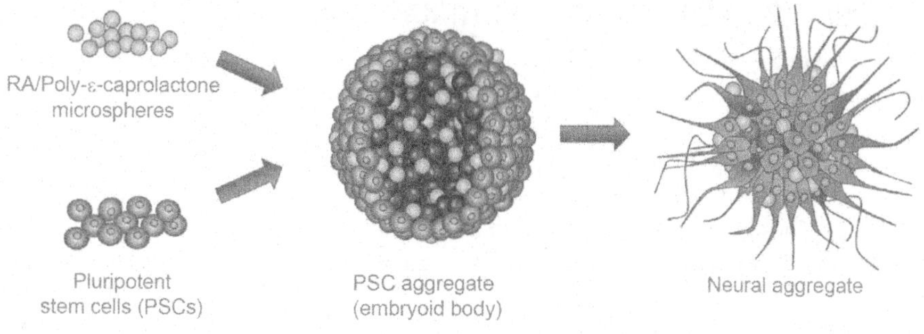

FIG. 4

Schematic of how to engineer neural tissue by combining drug releasing microspheres with pluripotent stem cells. In our 2015 Cellular and Molecular Bioengineering paper, we demonstrated that incorporating retinoic acid releasing microspheres in pluripotent stem cell aggregates resulted in the production of engineered neural tissues.

Reprinted with permission from Gomez JC, Edgar JM, Agbay AM, Bibault E, Montgomery A, Mohtaram NK, et al. Incorporation of retinoic acid releasing microspheres into pluripotent stem cell aggregates for inducing neuronal differentiation. Cell Mol Bioeng 2015;8(3):307–19.

My group also recently published a study that developed a novel method for engineering neural tissue from human-induced pluripotent stem cells as shown in Fig. 4 [106]. We developed PCL microspheres that could release retinoic acid over 28 days. These microspheres were then incorporated into aggregates of human-induced pluripotent stem cells, which then differentiated into neural tissue containing neurons. We are currently producing microspheres that deliver other small molecules as we hope to use different combinations of small molecule releasing microspheres to engineer neural tissue from different regions of the central nervous system. Our method provides a way for engineering neural tissue from stem cells that does not require the addition of soluble factors to culture media, which makes it easier to translate for clinical applications that require cell transplantation. We have also shown that the PCL microspheres can be used to deliver bioactive glial-derived neurotrophic factor as well [107]. Another group developed an injectable system using PCL microspheres that released NGF as a method for cell delivery [108]. Finally, a different study coated PCL microspheres with polydopamine as a way to get human neural stem cells to adhere to their surface [109]. These microspheres could also be decorated with adeno-associated virus (AAV) that could be used to transfect stem cells to influence their behavior.

The large body of work discussed in this section demonstrates the potential of novel PCL scaffolds for delivering physical and chemical cues for influencing stem cell behavior. While nanofiber scaffolds have been more extensively studied, the use of PCL microspheres as a means of controlling and delivering stem cells has been gaining in popularity. Overall, PCL remains an excellent biomaterial when it comes to applications in neural tissue engineering.

2.4 POLY (2-HYDROXYETHYL METHACRYLATE)

Poly (2-hydroxyethyl methacrylate) (PHEMA), a polymer made from $C_6H_{10}O_3$ monomers, forms stable hydrogels, which can be used for biomedical applications such as tissue engineering. For example, contact lens are often produced from PHEMA, indicating its nature as a biocompatible polymer [110,111]. Similar to other synthetic polymers, PHEMA can be modified to improve its bioactive properties [112]. These polymer hydrogels can also be fabricated to contain different types of topography, such as tubular structure that guide neurons during the regeneration process that occurs after spinal injury in a pre-clinical rat model [113]. Other methods of fabricating PHEMA into novel topography involve photolithography, which can be used to fabricate structures as well as to immobilize bioactive cues inside these scaffolds [114,115]. In 2004, a Biomaterials paper reported that mouse embryonic stem cells could be cultured on PHEMA scaffolds of varying porosity and this report surprisingly showed that coating such scaffolds with gelatin did not improve their ability to bind these stem cells [116]. However, more recent studies have functionalized PHEMA scaffolds with peptide sequences that promote adhesion of cells, including stem cells [117]. These features make PHEMA a desirable biomaterial for engineering neural tissue using stem cells.

Several versions of PHEMA scaffolds have been used for the repair of spinal cord injury. Such scaffolds can deliver human neural progenitors derived from the fetal spinal cord in combination with serotonin [118]. Most the cells migrated out of the scaffold and matured into neurons, but this engineered tissue did not promote functional recovery. A recent study showed that porous 3D PHEMA scaffolds modified to contain the peptide sequences RGDS and SIKVAVS supported the culture of both mesenchymal stem cells and neural stem cells [119]. For both types of cells, the scaffolds containing the RGDS sequence were better at supporting stem cell culture when compared to scaffolds functionalized with the SIKVAVS sequence. Another recent study from Dr. Schaffer's lab used PHEMA substrates to present different types of DNA ligands to create arrays consisting of types of stem cells through DNA-specific adhesion [120]. This system provides an interesting high-throughput platform for studying stem cell behavior by enabling generation of tissue arrays with different neural cell compositions. While there have been relatively few studies using PHEMA for neural tissue engineering using stem cells, it has a significant potential for such applications. PHEMA is quite stable as it has a slow degradation rate. It will also be discussed later in Section 3, which discusses stimulus-responsive materials.

2.5 POLYPYRROLE

Pyrrole, a five-membered ring with the formula C_4H_4NH, can be synthesized into a conducting polymer known as polypyrrole (Ppy) [121,122]. The conductive nature of Ppy made it a logical material choice for engineering neural probes [123,124] as well as for tissue engineering applications as electrical stimulation can improve neural cell functions [125–127]. Initial cell culture studies showed that Ppy was less suitable for olfactory cell culture compared to other polymers [128]. However, like other polymers, Ppy can be modified to contain sites for cell adhesion [129,130] as well as deliver drugs that enhance its bioactivity and biocompatibility [131]. These modified scaffolds are compatible with mature neural cell culture, suggesting that they could be suitable for neural stem cell culture [132]. Ppy scaffolds can be conjugated with bioactive NGF as a method for engineering neural tissue [133,134]. The NGF-Ppy scaffolds enhanced neurite extension of PC12 cells (a neural cancer cell line) at similar levels compared to treatment with soluble NGF. Further work from Dr. Christine Schmidt showed that micropatterned Ppy scaffolds promoted faster polarization of embryonic hippocampal neurons compared to scaffolds without patterning as seen in Fig. 5 [135].

In terms of compatibility with stem cell culture, Ppy scaffolds doped with laminin-derived peptides enhanced the differentiation of both human embryonic stem cells and rat neural stem cells into neurons [136]. Another study showed that Ppy scaffold

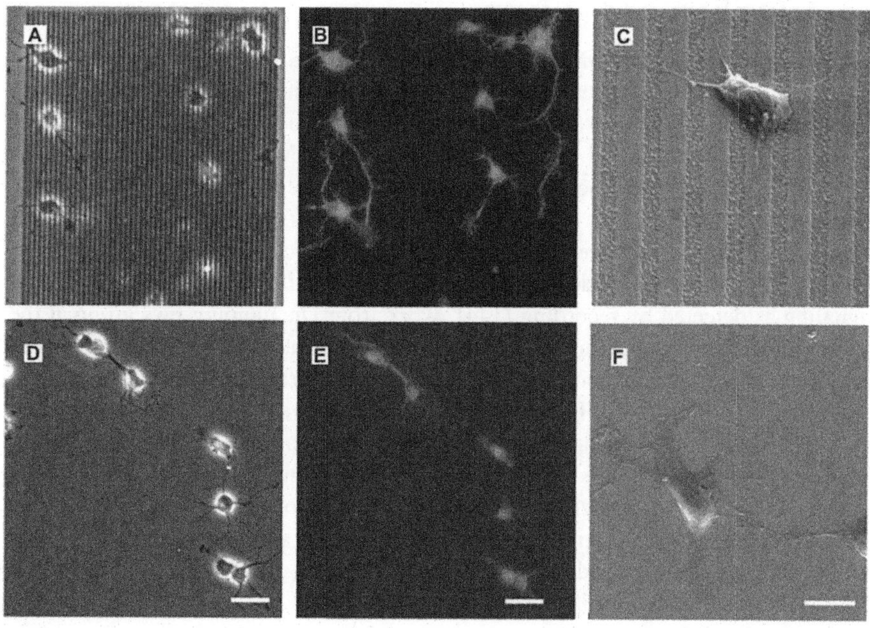

FIG. 5

Polypyrole scaffolds support neural cell culture. From left to right: phase-contrast, fluorescence, and SEM microscopy images, of hippocampal neurons on PPy. (A–C) Cells cultured on 2 μm wide and 200 nm deep PPy microchannels; (D–F) cells cultured on unmodified PPy. The *green* labeling (Alexa 488) corresponds to Tau-1 (axonal marker) immunostaining *in the electronic version, gray in the print version.* Cells polarized more readily on microchannels than on unmodified PPy. The scale bar is 20 μm (A,B,D,E) and 5 μm (C,F).

Reprinted with permission from Gomez N, Lee JY, Nickels JD, Schmidt CE. Micropatterned polypyrrole: a combination of electrical and topographical characteristics for the stimulation of cells. Adv Funct Mater 2007;17(10):1645–53.

doped with dodecylbenzenesulfonate supported rat fetal neural stem cell culture and then these scaffolds could be altered using electrical current to influence stem behavior [137]. Recent work demonstrated that electrical stimulation using Ppy scaffolds promotes differentiation of retinal progenitor cells in more mature neurons compared to unstimulated cells cultured under the same conditions [138]. Similar effects were observed regarding the differentiation of human neural stem cells as electrical stimulation promoted their differentiation into mainly mature neurons [139]. Finally, Ppy was combined with alginate to produce a hybrid hydrogel capable of supporting human mesenchymal stem cell differentiation into neural phenotypes [140]. Overall, this set of studies indicates one of the major benefits of using Ppy is the ability to use electrical stimulation to promote neuronal differentiation. The use of electrical cues provides an additional avenue when designing scaffolds for effective engineering of neural tissue from stem cells.

2.6 POLY(DIMETHYLSILOXANE)

Poly(dimethylsiloxane) (PDMS), which is classified as a silicone, can be used for a variety of biomaterial applications, including contact lens [141]. It has gained popularity as the material is used to fabricate different types of devices for studying neural stem cell behavior [142]. Initial studies used PDMS to fabricate microchips that could generate neural networks by differentiating neural stem cells into mature neurons [143]. Other studies have confirmed that neurons derived from mouse neural stem cells could be cocultured with primary mouse cortical neurons and when these networks were cultured inside of PDMS microchannels, they would produce electric signaling as detected by recording [144]. Work from Dr. Jianping Fu's group at the University of Michigan demonstrated that microposts made from PDMS possessing different levels of rigidity influenced the neuronal differentiation of human embryonic stem cells through the Hippo/Yap signaling pathways [145]. Other studies using PDMS have used microarchitecture to recreate the neuromuscular junction; however, this work used primary cells as opposed to deriving them from stem cells [146]. In addition to being used in microfluidic devices, macroporous 3D PDMS scaffolds can be fabricated for tissue engineering applications using stem cells—though such a strategy has not been applied to neural tissue engineering yet [147]. Overall, PDMS has significant applications in neural tissue engineering.

2.7 SELF-ASSEMBLING PEPTIDES

Chapter 5 discussed the use of protein-based biomaterials and their application to engineering neural tissue from stem cells in detail. It is a logical extension of this work to generate synthetic peptide sequences that can assemble into hydrogels for tissue engineering application as peptide synthesis avoids some of the sourcing issues associated with isolating naturally derived proteins [148]. Synthetic peptides are produced using chemical reactions by peptide synthesizers, avoiding the issue of having to obtain proteins from natural sources. These synthetic peptides are often designed so that they self-assemble into structures that can support cell growth. As discussed in other sections, specific peptide sequences can be added to such scaffolds to improve cell adhesion and modulate stem cell behavior though this section will focus on the types of peptides that can form hydrogels through self-assembly. This subject (self-assembling peptides for applications in neural tissue engineering) was reviewed recently by Koss and Unsworth [149]. An excellent review of peptide sequences used for different applications was written by Lampe and Heilshorn in 2012 [6]. The peptide sequences are often referred to by the set of one letter abbreviations for each amino acid of the sequence and a link to the table of these abbreviations is included in the resource section at the end of the chapter. For example, the commonly used RGD peptide consists of the amino acid sequence Arginine-Glycine-Aspartic Acid. Previous examples given in this chapter have used such peptide sequences to improve the biological properties of synthetic biomaterials. This section will focus on a specific class of peptide sequences that can self-assemble into hydrogels that support stem cell differentiation into neural phenotypes. These scaffolds can gel upon injection, making them attractive for clinical applications.

One of the most commonly used peptides that can self-assemble is called RADA16 as it consists of 16 repeats of the amino acid sequence RADA [150]. These self-assembling peptide scaffolds are available commercially under the name Puramatrix©. This peptide forms a β sheet structure, which can then hydrogen bond to otherβ sheets, generating a novel self-assembling nanofiber scaffold. Fig. 6 shows the RADA peptide structure along with other variants evaluated in the same study. These scaffolds were able to promote regeneration in a retinal injury model in hamsters, demonstrating their potential for applications in neural tissue engineering [151]. The Zhang group then evaluated such scaffolds for their ability to support mouse adult neural stem cell culture and differentiation into neural phenotypes [152]. This paper examined several RADA16 variants functionalized with sequences taken from extracellular matrix proteins and it found that motifs taken from bone morphogenic protein enhanced cell survival. These neural stem cells could also differentiate in mature neural phenotypes inside of such synthetic peptide scaffolds. Follow-up work showed that coating these peptide scaffolds with laminin improved neural stem cell survival and differentiation [153]. These scaffolds can also support long-term culture of neural stem cells for 150 days [154]. A different study found that such RADA16 scaffolds were more effective at supporting the culture of human fetal neural stem cells compared to other synthetic scaffolds [155].

Dr. Michael Fehling's research group, which is part of the University Health Network in Toronto, Ontario, has also characterized a different, novel self-assembling peptide for applications in delivering stem cells to damaged nervous system. This peptide consists of the sequence K2(QL)6 K2(QL6) and it was able to reduce glial scarring when used post spinal cord injury in rats [156]. Follow-up work showed that these peptide scaffolds could significantly increase the survival of neural progenitor cells delivered into the injured spinal cord while enhancing their ability to differentiate into mature neural phenotypes and integrate with the host tissue. Overall, this effect led to an improvement in functional recovery post injury [157]. Similar effects were observed in

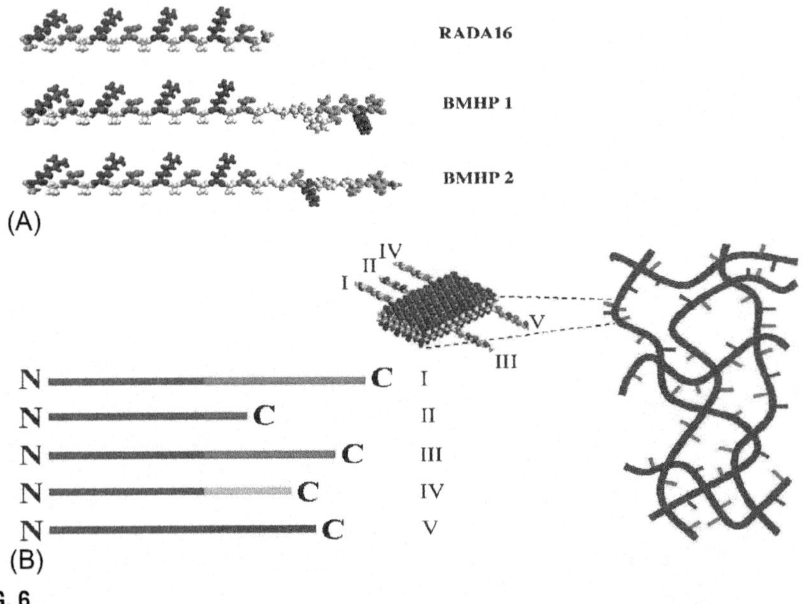

FIG. 6

Molecular and schematic models of the designer peptides and of the scaffolds. (A) Molecular models of RADA16, RADA16-Bone Marrow Homing Peptide 1 (BMHP1) and RADA16-Bone Marrow Homing Peptide 2 (BMHP2). RADA16 is an alternating16-residue peptide with basic arginine *(blue in the electronic version, gray in the print version)*, hydrophobic alanine *(white)*, and aspartic acid *(red in the electronic version, gray in the print version)*. These peptides self-assemble once exposed to physiological pH solutions or salt. The alanines of the RADA16 providing hydrophobic interaction are on one side of the peptide, and the arginines and aspartates form complementary ionic bonds on the other. The BMHP1 and BMHP2 motifs were directly extended from RADA16 with two glycine spacers and are composed of a lysine *(blue in the electronic version, gray in the print version)*, serine and threonine *(green in the electronic version, gray in the print version)*, and different hydrophobic *(white)* residues. Neutral polar residues are drawn in *green*. (B) Schematic models of several different functional motifs (different colored bars) could be extended from RADA16 *(blue bars in the electronic version, gray in the print version)* in order to design different peptides (I, II, III, IV, and V). They can be combined in different ratios. A schematic model of a self-assembling nanofiber scaffold with combinatorial motifs carrying different biological functions is shown.

Reprinted with permission from Gelain F, Bottai D, Vescovi A, Zhang S. Designer self-assembling peptide nanofiber scaffolds for adult mouse neural stem cell 3-dimensional cultures. PloS ONE 2006;1(1):e119.

more clinically relevant cervical injury models of spinal cord injury as well [158,159]. They recently published detailed protocols for using these scaffolds to deliver cells into their spinal cord injury models in the Journal of Visual Experiments [160].

Another study analyzed a mixture of synthetic peptides consisting of 200 residues in length with varying compositions to determine which peptide scaffold was optimal for delivering drugs to the central nervous system [161]. They found that increasing the alanine content of these hydrogels enabled better loading of hydrophobic molecules and these peptide-based scaffolds could deliver drugs in vivo to the damaged regions of the brain. Such work suggests that longer peptides can also be used to fabricate hydrogels for applications in neural tissue engineering. Other work has validated the use of a synthetic peptide mimetic of the neural cell adhesion molecule (NCAM) protein as a tool for influencing neural stem cell behavior [162]. This peptide enhanced proliferation of neural stem cells while inducing differentiation into oligodendrocytes. Directly injecting this peptide into the brain achieved similar effects. In another approach, synthetic proteoglycans were produced that could bind basic fibroblast growth factor [163]. These proteoglycans were then incorporated into the membranes of embryonic stem cells, where they became incorporated and could generate activation of the signaling pathways associated with fibroblast growth factor. Overall, peptide scaffolds provide flexibility and tunable bioactivity when used as scaffolds for engineering neural tissue.

2.8 OTHER POTENTIAL POLYMERS SUITABLE FOR APPLICATIONS IN NEURAL TISSUE ENGINEERING USING STEM CELLS

While the previous sections have discussed synthetic polymer scaffolds for directing stem cell differentiation into neural phenotypes, other synthetic polymers exist that have been evaluated for applications in neural tissue engineering that did not involve stem cells. The studies detailed in this section indicate the promise of other synthetic polymers and future work can explore their compatibility as substrates for stem cell culture and promoting their differentiation into mature neural phenotypes.

Poly (ethylene-*co*-vinylacetate) (PEVA), which is sometimes referred as (ethylene-co-vinylacetate) (EVA), consists of monomers of both ethylene and vinyl acetate. The ratio of these components can be altered and changing the ratio influences the properties of the resulting polymer. This polymer exhibits elastomeric behavior and its common applications include adhesives, being added to resins and biomedical applications. This polymer has been successfully used for applications in drug delivery to the brain, including the delivery of active growth factors and neurotransmitters [164–169]. It could be potentially used to fabricate drug releasing microspheres for promoting stem cell differentiation that could be incorporated into either natural or synthetic scaffolds.

Carbon nanotubes are another interesting biomaterial for further research, as the unique nature of carbon being able to form four covalent bonds makes it a common component of organic compounds [170]. The cylindrical structure of carbon nanotubes gives them excellent mechanical strength, while they also exhibit excellent conductivity, which is a desirable feature when engineering materials for neural tissue engineering applications [171]. Thus, carbon nanotubes have been added to different types of biomaterial scaffolds to improve their properties [172,173]. For example, synthetic PLL electrospun scaffolds impregnated with carbon nanotubes supported the differentiation of mouse embryonic stem cells into neural lineages [174]. Another approach combined collagen with carbon nanotubes to generate scaffolds that promoted the rapid differentiation of human embryonic stem cells into ectoderm cells that were nesting positive [175]. The amount of carbon nanotubes added can also modulate the stiffness of collagen hydrogels as well [176]. Overall, these properties and examples illustrate the potential of carbon nanotubes for applications in neural tissue engineering.

3 THE DEVELOPMENT OF "SMART" BIOMATERIALS THAT RESPOND TO STIMULI

The rest of the chapter has dealt directly with synthetic biomaterials that have been evaluated for their ability to direct stem cells to form neural tissue. Often these materials were functionalized with additional cues to enhance their inherent properties using various types of chemistry. In addition to these studies, other research groups have developed various ways of producing materials that respond to changes in their environment induced by cells. These responsive materials are often termed "smart" because of their ability to change to their properties in response to stimuli. Such materials have the potential to expand further the possibilities when using synthetic scaffolds to engineer neural tissue from stem cells. For example, PHEMA hydrogels can also be formulated to change their mechanical properties in response to changes in temperature [177]. Other synthetic hydrogels can be produced that respond to changes in pH, which provides another avenue for controlling the properties of a material [178]. Another interesting avenue for generating smart materials is to use electrically conducting polymers [122]. For example, dexamethasone can be released from Ppy scaffolds through electrical stimulation [131]. Another type of smart hydrogel responds to changes in light—in this case, releasing drugs when stimulated with near infrared light [179]. In fact, a recent study combined these different properties into one material that could respond to changes in pH, temperature, and light by changing between two different shapes [180]. Magnetic components can also be incorporated into hydrogels for applications in drug delivery [181]. These properties can all be used to encapsulate cells or trigger changes in the microenvironment to promote desired differentiation of stem cells. These methods result in the production of stimulus-responsive hydrogels, which could potentially be translated for neural tissue engineering applications.

Additionally, this chapter has discussed the use of electrospun nanofibers as scaffolds for directing stem cell differentiation into neurons. These scaffolds can also be formulated to be responsive to stimuli [182]. For example, such nanofibers can be fabricated to respond to changes of pH based on the chemistry of the polymer used [183,184]. These nanofibers can release drugs or change their fiber morphology in response to these pH changes. The pH levels in the body fluctuate and cells can alter the pH of their environment due to changes in metabolism, and this property can be exploited when designing biomaterial scaffolds as a trigger for releasing drugs or changing hydrogel properties. Thermoresponsive polymers can be spun in electrospun scaffolds, which can release drugs in response to changes in temperature [185,186]. These different types of smart biomaterials provide an interesting avenue for generating changes in scaffold topography as well as for controlling the release of bioactive molecules. Several of these systems will be explored in depth in Chapter 7, which details different types of drug delivery systems. These technologies serve as an excellent starting point for future work adapting these responsive materials as a tool for controlling stem cell behavior.

4 FUTURE DIRECTIONS

This chapter has detailed the wide variety of synthetic biomaterials that have been used as substrates for neural tissue engineering using stem cells. These materials have been used to synthesize scaffolds with novel topographies, a range of mechanical properties, and the presentation of unique biochemical cues. Their ability to be modified opens up a wide range of potential with regard to tissue engineering, including the development of stimulus-responsive materials as detailed in Section 3. Additionally, several synthetic polymers have yet to be evaluated for their compatibility with stem cells for neural applications as discussed in Section 2.8. Both avenues provide future directions for generating functional neural tissues from stem cells.

ADDITIONAL RESOURCES

LINK TO THE CHART SHOWING THE SINGLE LETTER ABBREVIATIONS FOR EACH AMINO ACID

https://www.ncbi.nlm.nih.gov/Class/MLACourse/Modules/MolBioReview/iupac_aa_abbreviations.html

LINKS TO WEB RESOURCES ABOUT BIOMATERIALS

Society for Biomaterials: https://www.biomaterials.org/

Canadian Society for Biomaterials: https://biomaterials.ca/

Tissue Engineering and Regenerative Medicine International Society: https://www.termis.org/

Biomedical Engineering Society: http://www.bmes.org/

Nature's website on Biomaterial Research: http://www.nature.com/subjects/biomaterials

Biomaterials—the journal: http://www.journals.elsevier.com/biomaterials/

Tissue Engineering—the journal: http://www.liebertpub.com/overview/tissue-engineering-parts-a-b-and-c/595/

TEXTBOOKS ABOUT BIOMATERIALS FOR FURTHER READING

Biomaterials Science: An Introduction to Materials in Medicine (Third Edition). Edited by Buddy Ratner, Allan S. Hoffman, Fredrick J. Schoen, and Jack E. Lemons. 3rd edition. Massachusetts: Elsevier. 2013.

Temenoff, J.S. and Mikos, A.G. Biomaterials: The Intersection of Biology and Materials Science. 1st edition. London: Pearson Science. 2008.

Williams, D.F. Essential Biomaterial Science (Cambridge Texts in Biomedical Engineering). 1st edition. Cambridge: Cambridge University Press 2014.

TEXTBOOKS ABOUT TISSUE ENGINEERING FOR FURTHER READING

Saltzman, M. Tissue Engineering: Engineering Principles for the Design of Replacement Organs and Tissues. Oxford University Press. 1st edition. 2004.

Principles of Tissue Engineering. Edited by Robert Lanza, Robert Langer, and Joseph Vacanti. Academic Press. Fourth edition. 2013.

REFERENCES

[1] Carlini AS, Adamiak L, Gianneschi NC. Biosynthetic polymers as functional materials. Macromolecules 2016;49(12):4379–94.

[2] Mansouri N. The influence of topography on tissue engineering perspective. Mater Sci Eng C 2016;61:906–21.

[3] Elliott Donaghue I, Tam R, Sefton MV, Shoichet MS. Cell and biomolecule delivery for tissue repair and regeneration in the central nervous system. J Control Release 2014;190:219–27.

[4] Cunha C, Panseri S, Antonini S. Emerging nanotechnology approaches in tissue engineering for peripheral nerve regeneration. Nanomed Nanotechnol Biol Med 2011;7(1):50–9.

[5] Guvendiren M, Burdick JA. Engineering synthetic hydrogel microenvironments to instruct stem cells. Curr Opin Biotechnol 2013;24(5):841–6.

[6] Lampe KJ, Heilshorn SC. Building stem cell niches from the molecule up through engineered peptide materials. Neurosci Lett 2012;519(2):138–46.

[7] Joglekar M, Trewyn BG. Polymer-based stimuli-responsive nanosystems for biomedical applications. Biotechnol J 2013;8(8):931–45.

[8] Singh NK, Lee DS. In situ gelling pH-and temperature-sensitive biodegradable block copolymer hydrogels for drug delivery. J Control Release 2014;193:214–27.

[9] Agbay A, Edgar JM, Robinson M, Styan T, Wilson K, Schroll J, et al. Biomaterial strategies for delivering stem cells as a treatment for spinal cord injury. Cells Tissues Organs 2016;202(1–2):42–51.

[10] Khalf A, Madihally SV. Recent advances in multiaxial electrospinning for drug delivery. Eur J Pharm Biopharm 2016;112:1–17.

[11] Chen N, Tian L, He L, Ramakrishna S. Nanobiomaterials for neural regeneration. Neural Regen Res 2016;11(9):1372.

[12] Willerth SM. Melt electrospinning in tissue engineering. In: Kny E, Uyar T, Khenoussi N, editors. Electrospun materials for tissue engineering and biomedical applications: research, design and commercialization. Amsterdam, Netherlands: Elsevier; 2017.

[13] Dalton PD, Joergensen NT, Groll J, Moeller M. Patterned melt electrospun substrates for tissue engineering. Biomed Mater 2008;3(3):034109.

[14] Brown TD, Dalton PD, Hutmacher DW. Direct writing by way of melt electrospinning. Adv Mater 2011;23(47):5651–7.

[15] Hutmacher DW, Dalton PD. Melt electrospinning. Chem Asian J 2011;6(1):44–56.

[16] Nagy ZK, Balogh A, Dravavolgyi G, Ferguson J, Pataki H, Vajna B, et al. Solvent-free melt electrospinning for preparation of fast dissolving drug delivery system and comparison with solvent-based electrospun and melt extruded systems. J Pharm Sci 2013;102(2):508–17.

[17] Balogh A, Dravavolgyi G, Farago K, Farkas A, Vigh T, Soti PL, et al. Plasticized drug-loaded melt electrospun polymer mats: characterization, thermal degradation, and release kinetics. J Pharm Sci 2014;103(4):1278–87.

[18] Balogh A, Farkas B, Farago K, Farkas A, Wagner I, Van Assche I, et al. Melt-blown and electrospun drug-loaded polymer fiber mats for dissolution enhancement: a comparative study. J Pharm Sci 2015;104(5):1767–76.

[19] Kim J, Kong YP, Niedzielski SM, Singh RK, Putnam AJ, Shikanov A. Characterization of the crosslinking kinetics of multi-arm poly (ethylene glycol) hydrogels formed via Michael-type addition. Soft Matter 2016;12(7):2076–85.

[20] Ivirico JLE, Cruz DMG, Monrós MCA, Martínez-Ramos C, Pradas MM. Synthesis and properties of caprolactone and ethylene glycol copolymers for neural regeneration. J Mater Sci Mater Med 2012;23(7):1605–17.

[21] Escudero-Castellanos A, Ocampo-García BE, Domínguez-García MV, Flores-Estrada J, Flores-Merino MV. Hydrogels based on poly (ethylene glycol) as scaffolds for tissue engineering application: biocompatibility assessment and effect of the sterilization process. J Mater Sci Mater Med 2016;27(12):176.

[22] Tirelli N, Lutolf M, Napoli A, Hubbell J. Poly (ethylene glycol) block copolymers. Rev Mol Biotechnol 2002;90(1):3–15.

[23] Sargeant TD, Desai AP, Banerjee S, Agawu A, Stopek JB. An in situ forming collagen–PEG hydrogel for tissue regeneration. Acta Biomater 2012;8(1):124–32.

[24] Tseng TC, Tao L, Hsieh FY, Wei Y, Chiu IM, Hsu S. An injectable, self-healing hydrogel to repair the central nervous system. Adv Mater 2015;27(23):3518–24.

[25] Dai W, Belt J, Saltzman WM. Cell-binding peptides conjugated to poly (ethylene glycol) promote neural cell aggregation. Nat Biotechnol 1994;12(8):797–801.

[26] Burdick JA, Ward M, Liang E, Young MJ, Langer R. Stimulation of neurite outgrowth by neurotrophins delivered from degradable hydrogels. Biomaterials 2006;27(3):452–9.

[27] Royce Hynes S, McGregor LM, Ford Rauch M, Lavik EB. Photopolymerized poly (ethylene glycol)/poly (L-lysine) hydrogels for the delivery of neural progenitor cells. J Biomater Sci Polym Ed 2007;18(8):1017–30.

[28] Hynes SR, Rauch MF, Bertram JP, Lavik EB. A library of tunable poly (ethylene glycol)/poly (L-lysine) hydrogels to investigate the material cues that influence neural stem cell differentiation. J Biomed Mater Res A 2009;89(2):499–509.

[29] Rauch MF, Michaud M, Xu H, Madri JA, Lavik EB. Co-culture of primary neural progenitor and endothelial cells in a macroporous gel promotes stable vascular networks in vivo. J Biomater Sci Polym Ed 2008;19(11):1469–85.

[30] Cai L, Lu J, Sheen V, Wang S. Optimal poly (L-lysine) grafting density in hydrogels for promoting neural progenitor cell functions. Biomacromolecules 2012;13(5):1663–74.

[31] Franco C, Price J, West J. Development and optimization of a dual-photoinitiator, emulsion-based technique for rapid generation of cell-laden hydrogel microspheres. Acta Biomater 2011;7(9):3267–76.

[32] Pellett S, Schwartz MP, Tepp WH, Josephson R, Scherf JM, Pier CL, et al. Human induced pluripotent stem cell derived neuronal cells cultured on chemically-defined hydrogels for sensitive in vitro detection of botulinum neurotoxin. Sci Rep 2015;5.

[33] Roccio M, Gobaa S, Lutolf MP. High-throughput clonal analysis of neural stem cells in microarrayed artificial niches. Integr Biol 2012;4(4):391–400.

[34] Joshi R, Tavana H, editors. Microengineered embryonic stem cells niche to induce neural differentiation. 2015 37th Annual international conference of the IEEE Engineering in Medicine and Biology Society (EMBC); IEEE; 2015.

[35] Li X, Liu X, Cui L, Brunson C, Zhao W, Bhat NR, et al. Engineering an in situ cross-linkable hydrogel for enhanced remyelination. FASEB J 2013;27(3):1127–36.

[36] Li X, Liu X, Zhang N, Wen X. Engineering in situ cross-linkable and neurocompatible hydrogels. J Neurotrauma 2014;31(16):1431–8.

[37] Aurand ER, Wagner JL, Shandas R, Bjugstad KB. Hydrogel formulation determines cell fate of fetal and adult neural progenitor cells. Stem Cell Res 2014;12(1):11–23.

[38] Namba RM, Cole AA, Bjugstad KB, Mahoney MJ. Development of porous PEG hydrogels that enable efficient, uniform cell-seeding and permit early neural process extension. Acta Biomater 2009;5(6):1884–97.

[39] Lampe KJ, Bjugstad KB, Mahoney MJ. Impact of degradable macromer content in a poly (ethylene glycol) hydrogel on neural cell metabolic activity, redox state, proliferation, and differentiation. Tissue Eng A 2010;16(6):1857–66.

[40] Mooney R, Haeger S, Lawal R, Mason M, Shrestha N, Laperle A, et al. Control of neural cell composition in poly (ethylene glycol) hydrogel culture with soluble factors. Tissue Eng A 2011;17(21–22):2805–15.

[41] Scott R, Marquardt L, Willits RK. Characterization of poly(ethylene glycol) gels with added collagen for neural tissue engineering. J Biomed Mater Res A 2010;93A(3):817–23.

[42] Zhou W, Blewitt M, Hobgood A, Willits RK. Comparison of neurite growth in three dimensional natural and synthetic hydrogels. J Biomater Sci Polym Ed 2013;24(3):301–14.

[43] Lawrence PB, Price JL. How PEGylation influences protein conformational stability. Curr Opin Chem Biol 2016;34:88–94.

[44] Sarig-Nadir O, Seliktar D. Compositional alterations of fibrin-based materials for regulating in vitro neural outgrowth. Tissue Eng A 2008;14(3):401–11.

[45] Pepinsky RB, Walus L, Shao Z, Ji B, Gu S, Sun Y, et al. Production of a PEGylated Fab' of the anti-LINGO-1 Li33 antibody and assessment of its biochemical and functional properties in vitro and in a rat model of Remyelination. Bioconjug Chem 2011;22(2):200–10.

[46] Stukel J, Thompson S, Simon L, Willits R. Polyethlyene glycol microgels to deliver bioactive nerve growth factor. J Biomed Mater Res A 2015;103(2):604–13.

[47] Roam JL, Nguyen PK, Elbert DL. Controlled release and gradient formation of human glial-cell derived neurotrophic factor from heparinated poly (ethylene glycol) microsphere-based scaffolds. Biomaterials 2014;35(24):6473–81.

[48] Roam JL, Yan Y, Nguyen PK, Kinstlinger IS, Leuchter MK, Hunter DA, et al. A modular, plasmin-sensitive, clickable poly (ethylene glycol)-heparin-laminin microsphere system for establishing growth factor gradients in nerve guidance conduits. Biomaterials 2015;72:112–24.

[49] Curley JL, Jennings SR, Moore MJ. Fabrication of micropatterned hydrogels for neural culture systems using dynamic mask projection photolithography. J Vis Exp 2011;(48)e2636-e.

[50] Evangelista MS, Perez M, Salibian AA, Hassan JM, Darcy S, Paydar KZ, et al. Single-lumen and multi-lumen poly (ethylene glycol) nerve conduits fabricated by stereolithography for peripheral nerve regeneration in vivo. J Reconstr Microsurg 2015;31(05):327–35.

[51] Meinhardt A, Eberle D, Tazaki A, Ranga A, Niesche M, Wilsch-Bräuninger M, et al. 3D reconstitution of the patterned neural tube from embryonic stem cells. Stem Cell Rep 2014;3(6):987–99.

[52] Xu Y, Kim CS, Saylor DM, Koo D. Polymer degradation and drug delivery in PLGA-based drug–polymer applications: a review of experiments and theories. J Biomed Mater Res B Appl Biomater 2016.

[53] Cai Q, Wang L, Deng G, Liu J, Chen Q, Chen Z. Systemic delivery to central nervous system by engineered PLGA nanoparticles. Am J Transl Res 2016;8(2):749.

[54] Tseng Y-Y, Liu S-J. Nanofibers used for the delivery of analgesics. Nanomedicine 2015;10(11):1785–800.

[55] Lavik E, Teng YD, Snyder E, Langer R. Seeding neural stem cells on scaffolds of PGA, PLA, and their copolymers. Neural Stem Cells: Methods Protoc 2002;89–96.

[56] Levenberg S, Huang NF, Lavik E, Rogers AB, Itskovitz-Eldor J, Langer R. Differentiation of human embryonic stem cells on three-dimensional polymer scaffolds. Proc Natl Acad Sci 2003;100(22):12741–6.

[57] Levenberg S, Burdick JA, Kraehenbuehl T, Langer R. Neurotrophin-induced differentiation of human embryonic stem cells on three-dimensional polymeric scaffolds. Tissue Eng 2005;11(3–4):506–12.

[58] Hou S-Y, Zhang H-Y, Quan D-P, Liu X-L, Zhu J-K. Tissue-engineered peripheral nerve grafting by differentiated bone marrow stromal cells. Neuroscience 2006;140(1):101–10.

[59] Bhang SH, Lim JS, Choi CY, Kwon YK, Kim B-S. The behavior of neural stem cells on biodegradable synthetic polymers. J Biomater Sci Polym Ed 2007;18(2):223–39.

[60] Rooney GE, Moran C, McMahon SS, Ritter T, Maenz M, Flügel A, et al. Gene-modified mesenchymal stem cells express functionally active nerve growth factor on an engineered poly lactic glycolic acid (PLGA) substrate. Tissue Eng A 2008;14(5):681–90.

[61] Xiong Y, Zeng Y-S, Zeng C-G, B-l D, He L-M, Quan D-P, et al. Synaptic transmission of neural stem cells seeded in 3-dimensional PLGA scaffolds. Biomaterials 2009;30(22):3711–22.

[62] Rauch MF, Hynes SR, Bertram J, Redmond A, Robinson R, Williams C, et al. Engineering angiogenesis following spinal cord injury: a coculture of neural progenitor and endothelial cells in a degradable polymer implant leads to an increase in vessel density and formation of the blood–spinal cord barrier. Eur J Neurosci 2009;29(1):132–45.

[63] Olson HE, Rooney GE, Gross L, Nesbitt JJ, Galvin KE, Knight A, et al. Neural stem cell–and Schwann cell–loaded biodegradable polymer scaffolds support axonal regeneration in the transected spinal cord. Tissue Eng A 2009;15(7):1797–805.

[64] Yu D, Neeley WL, Pritchard CD, Slotkin JR, Woodard EJ, Langer R, et al. Blockade of peroxynitrite-induced neural stem cell death in the acutely injured spinal cord by drug-releasing polymer. Stem Cells 2009;27(5):1212–22.

[65] Kim BG, Kang YM, Phi JH, Kim YH, Hwang DH, Choi JY, et al. Implantation of polymer scaffolds seeded with neural stem cells in a canine spinal cord injury model. Cytotherapy 2010;12(6):841–5.

[66] Khang G, Kim HL, Hong M, Lee D. Neurogenesis of bone marrow-derived mesenchymal stem cells onto β-mercaptoethanol-loaded PLGA film. Cell Tissue Res 2012;347(3):713–24.

[67] Lowry N, Goderie SK, Lederman P, Charniga C, Gooch MR, Gracey KD, et al. The effect of long-term release of Shh from implanted biodegradable microspheres on recovery from spinal cord injury in mice. Biomaterials 2012;33(10):2892–901.

[68] Li XX, Hou S, Feng XZ, Yu Y, Ma JJ, Li LY. Patterning of neural stem cells on poly(lactic-co-glycolic acid) film modified by hydrophobin. Colloids Surf B-Biointerfaces 2009;74(1):370–4.

[69] Zhang Y-q, He L-m, Xing B, Zeng X, Zeng C-g, Zhang W, et al. Neurotrophin-3 gene-modified Schwann cells promote TrkC gene-modified mesenchymal stem cells to differentiate into neuron-like cells in poly (lactic-acid-co-glycolic acid) multiple-channel conduit. Cells Tissues Organs 2011;195(4):313–22.

[70] Yang K, Park E, Lee JS, Kim IS, Hong K, Park KI, et al. Biodegradable nanotopography combined with neurotrophic signals enhances contact guidance and neuronal differentiation of human neural stem cells. Macromol Biosci 2015;15(10):1348–56.

[71] Worthington KS, Wiley LA, Guymon CA, Salem AK, Tucker BA. Differentiation of induced pluripotent stem cells to neural retinal precursor cells on porous poly-lactic-co-glycolic acid scaffolds. J Ocul Pharmacol Ther 2016;32(5):310–6.

[72] Nkansah MK, Tzeng SY, Holdt AM, Lavik EB. Poly (lactic-co-glycolic acid) nanospheres and microspheres for short-and long-term delivery of bioactive ciliary neurotrophic factor. Biotechnol Bioeng 2008;100(5):1010–9.

[73] Nojehdehian H, Moztarzadeh F, Baharvand H, Nazarian H, Tahriri M. Preparation and surface characterization of poly-L-lysine-coated PLGA microsphere scaffolds containing retinoic acid for nerve tissue engineering: in vitro study. Colloids Surf B: Biointerfaces 2009;73(1):23–9.

[74] Nojehdehian H, Moztarzadeh F, Baharvand H, Mehrjerdi NZ, Nazarian H, Tahriri M. Effect of poly-L-lysine coating on retinoic acid-loaded PLGA microspheres in the differentiation of carcinoma stem cells into neural cells. Int J Artif Organs 2010;33(10):721–30.

[75] Bible E, Chau DYS, Alexander MR, Price J, Shakesheff KM, Modo M. The support of neural stem cells transplanted into stroke-induced brain cavities by PLGA particles. Biomaterials 2009;30(16):2985–94.

[76] Bible E, Qutachi O, Chau DYS, Alexander MR, Shakesheff KM, Modo M. Neo-vascularization of the stroke cavity by implantation of human neural stem cells on VEGF-releasing PLGA microparticles. Biomaterials 2012;33(30):7435-+.

[77] Wang Y, Wei YT, Zu ZH, Ju RK, Guo MY, Wang XM, et al. Combination of hyaluronic acid hydrogel scaffold and PLGA microspheres for supporting survival of neural stem cells. Pharm Res 2011;28(6):1406–14.

[78] Kim H, Zahir T, Tator CH, Shoichet MS. Effects of dibutyryl cyclic-AMP on survival and neuronal differentiation of neural stem/progenitor cells transplanted into spinal cord injured rats. PLoS ONE 2011;6(6):e21744.

[79] Xiong Y, Zhu JX, Fang ZY, Zeng CG, Zhang C, Qi GL, et al. Coseeded Schwann cells myelinate neurites from differentiated neural stem cells in neurotrophin-3-loaded PLGA carriers. Int J Nanomedicine 2012;7:1977–89.

[80] Gujral C, Minagawa Y, Fujimoto K, Kitano H, Nakaji-Hirabayashi T. Biodegradable microparticles for strictly regulating the release of neurotrophic factors. J Control Release 2013;168(3):307–16.

[81] Dash TK, Konkimalla VB. Poly-small je, Ukrainian-caprolactone based formulations for drug delivery and tissue engineering: a review. J Control Release 2012;158(1):15–33.

[82] Labet M, Thielemans W. Synthesis of polycaprolactone: a review. Chem Soc Rev 2009;38(12):3484–504.

[83] Villarreal-Gómez LJ, Cornejo-Bravo JM, Vera-Graziano R, Grande D. Electrospinning as a powerful technique for biomedical applications: a critically selected survey. J Biomater Sci Polym Ed 2016;27(2):157–76.

[84] Nisbet D, Yu L, Zahir T, Forsythe JS, Shoichet M. Characterization of neural stem cells on electrospun poly (ε-caprolactone) submicron scaffolds: evaluating their potential in neural tissue engineering. J Biomater Sci Polym Ed 2008;19(5):623–34.

[85] Xie JW, Willerth SM, Li XR, Macewan MR, Rader A, Sakiyama-Elbert SE, et al. The differentiation of embryonic stem cells seeded on electrospun nanofibers into neural lineages. Biomaterials 2009;30(3):354–62.

[86] Mohtaram NK, Ko J, Montgomery A, Carlson M, Sun L, Wong A, et al. Multifunctional electrospun scaffolds for promoting neuronal differentiation of induced pluripotent stem cells. J Biomater Tissue Eng 2014;4(11):906–14.

[87] Mohtaram N, Ko J, Agbay A, Rattray D, Neill P, Rajwani A, et al. Development of a glial cell-derived neurotrophic factor-releasing artificial dura for neural tissue engineering applications. J Mater Chem B 2015;3(40):7974–85.

[88] Ko J, Mohtaram NK, Ahmed F, Montgomery A, Carlson M, Lee PC, et al. Fabrication of poly (−caprolactone) microfiber scaffolds with varying topography and mechanical properties for stem cell-based tissue engineering applications. J Biomater Sci Polym Ed 2014;25(1):1–17.

[89] Mohtaram NK, Ko J, King C, Sun L, Muller N, Jun MB, et al. Electrospun biomaterial scaffolds with varied topographies for neuronal differentiation of human-induced pluripotent stem cells. J Biomed Mater Res A 2014;103(8):2591–601.

[90] Mahairaki V, Lim SH, Christopherson GT, Xu L, Nasonkin I, Yu C, et al. Nanofiber matrices promote the neuronal differentiation of human embryonic stem cell-derived neural precursors in vitro. Tissue Eng A 2010;17(5–6):855–63.

[91] Czeisler C, Short A, Stocker B, Cronin J, Lannutti J, Winter J, et al. Surface topography during neural stem cell differentiation regulates cell migration and cell morphology. FASEB J 2016;30(1 Supplement):160.5.

[92] Horne MK, Nisbet DR, Forsythe JS, Parish CL. Three-dimensional Nanofibrous scaffolds incorporating immobilized BDNF promote proliferation and differentiation of cortical neural stem cells. Stem Cells Dev 2010;19(6):843–52.

[93] Zhou K, Thouas G, Bernard C, Forsythe JS. 3D presentation of a neurotrophic factor for the regulation of neural progenitor cells. Nanomedicine 2014;9(8):1239–51.

[94] Jiang X, Cao HQ, Shi LY, Ng SY, Stanton LW, Chew SY. Nanofiber topography and sustained biochemical signaling enhance human mesenchymal stem cell neural commitment. Acta Biomater 2012;8(3):1290–302.

[95] Wang TY, Bruggeman KA, Sheean RK, Turner BJ, Nisbet DR, Parish CL. Characterization of the stability and bio-functionality of tethered proteins on bioengineered scaffolds: implications for stem cell biology and tissue repair. J Biol Chem 2014;289(21):15044–51.

[96] Teo BKK, Tan G-DS, Yim EK. The synergistic effect of nanotopography and sustained dual release of hydrophobic and hydrophilic neurotrophic factors on human mesenchymal stem cell neuronal lineage commitment. Tissue Eng A 2014;20(15–16):2151–61.

[97] Bechara S, Wadman L, Popat KC. Electroconductive polymeric nanowire templates facilitates in vitro C17. 2 neural stem cell line adhesion, proliferation and differentiation. Acta Biomater 2011;7(7):2892–901.

[98] Çapkın M, Çakmak S, Kurt FÖ, Gümüşderelioğlu M, Şen BH, Türk BT, et al. Random/aligned electrospun PCL/PCL-collagen nanofibrous membranes: comparison of neural differentiation of rat AdMSCs and BMSCs. Biomed Mater 2012;7(4):045013.

[99] Li Y, Ceylan M, Shrestha B, Wang H, QR L, Asmatulu R, et al. Nanofibers support oligodendrocyte precursor cell growth and function as a neuron-free model for myelination study. Biomacromolecules 2013;15(1):319–26.

[100] Lawley E, Baranov P, Young M. Hybrid vitronectin-mimicking polycaprolactone scaffolds for human retinal progenitor cell differentiation and transplantation. J Biomater Appl 2015;29(6):894–902.
[101] Jahani H, Jalilian FA, Wu CY, Kaviani S, Soleimani M, Abassi N, et al. Controlled surface morphology and hydrophilicity of polycaprolactone toward selective differentiation of mesenchymal stem cells to neural like cells. J Biomed Mater Res A 2015;103(5):1875–81.
[102] Low WC, Rujitanaroj P-O, Lee D-K, Messersmith PB, Stanton LW, Goh E, et al. Nanofibrous scaffold-mediated REST knockdown to enhance neuronal differentiation of stem cells. Biomaterials 2013;34(14):3581–90.
[103] Diao HJ, Low WC, QR L, Chew SY. Topographical effects on fiber-mediated microRNA delivery to control oligodendroglial precursor cells development. Biomaterials 2015;70:105–14.
[104] Hwang DH, Kim HM, Kang YM, Joo IS, Cho C-S, Yoon B-W, et al. Combination of multifaceted strategies to maximize the therapeutic benefits of neural stem cell transplantation for spinal cord repair. Cell Transplant 2011;20(9):1361–79.
[105] Kim D-Y, Choi Y-S, Kim S-E, Lee J-H, Kim S-M, Kim Y-J, et al. In vivo effects of adipose-derived stem cells in inducing neuronal regeneration in Sprague-Dawley rats undergoing nerve defect bridged with polycaprolactone nanotubes. J Korean Med Sci 2014;29(Suppl 3):S183-S92.
[106] Gomez JC, Edgar JM, Agbay AM, Bibault E, Montgomery A, Mohtaram NK, et al. Incorporation of retinoic acid releasing microspheres into pluripotent stem cell aggregates for inducing neuronal differentiation. Cell Mol Bioeng 2015;8(3):307–19.
[107] Agbay A, Mohtaram NK, Willerth SM. Controlled release of glial cell line-derived neurotrophic factor from poly(epsilon-caprolactone) microspheres. Drug Deliv Transl Res 2014;4(2):159–70.
[108] Kim SY, Hwang J-Y, Shin US. Preparation of nano/macroporous polycaprolactone microspheres for an injectable cell delivery system using room temperature ionic liquid and camphene. J Colloid Interface Sci 2016;465:18–25.
[109] Kim M, Kim JS, Lee H, Jang JH. Polydopamine-decorated sticky, water-friendly, biodegradable Polycaprolactone cell carriers. Macromol Biosci 2016;16(5):738–47.
[110] Chirila TV. Melanized poly (HEMA) hydrogels: basic research and potential use. J Biomater Appl 1993;8(2):106–44.
[111] Chirila TV. An overview of the development of artificial corneas with porous skirts and the use of PHEMA for such an application. Biomaterials 2001;22(24):3311–7.
[112] Oral E, Peppas NA. Hydrophilic molecularly imprinted poly (hydroxyethyl-methacrylate) polymers. J Biomed Mater Res A 2006;78(1):205–10.
[113] Tsai EC, Dalton PD, Shoichet MS, Tator CH. Synthetic hydrogel guidance channels facilitate regeneration of adult rat brainstem motor axons after complete spinal cord transection. J Neurotrauma 2004;21(6):789–804.
[114] Bryant SJ, Cuy JL, Hauch KD, Ratner BD. Photo-patterning of porous hydrogels for tissue engineering. Biomaterials 2007;28(19):2978–86.
[115] Hu Y, You J-O, Aizenberg J. Micropatterned hydrogel surface with high-aspect-ratio features for cell guidance and tissue growth. ACS Appl Mater Interfaces 2016;8(34):21939–45.
[116] Horák D, Kroupová J, Šlouf M, Dvořák P. Poly (2-hydroxyethyl methacrylate)-based slabs as a mouse embryonic stem cell support. Biomaterials 2004;25(22):5249–60.
[117] Kubinová Š, Horák D, Vaněček V, Plichta Z, Proks V, Syková E. The use of new surface-modified poly (2-hydroxyethyl methacrylate) hydrogels in tissue engineering: Treatment of the surface with fibronectin subunits versus ac-CGGASIKVAVS-OH, cysteine, and 2-mercaptoethanol modification. J Biomed Mater Res A 2014;102(7):2315–23.

[118] Růžička J, Romanyuk N, Hejčl A, Vetrik M, Hrubý M, Cocks G, et al. Treating spinal cord injury in rats with a combination of human fetal neural stem cells and hydrogels modified with serotonin. Acta Neurobiol Exp 2013;73:102–15.

[119] Macková H, Plichta Z, Proks V, Kotelnikov I, Kučka J, Hlídková H, et al. RGDS-and SIKVAVS-modified Superporous poly (2-hydroxyethyl methacrylate) scaffolds for tissue engineering applications. Macromol Biosci 2016;16(11):1621–31.

[120] Chen S, Bremer AW, Scheideler OJ, Na YS, Todhunter ME, Hsiao S, et al. Interrogating cellular fate decisions with high-throughput arrays of multiplexed cellular communities. Nat Commun 2016;7.

[121] Green RA, Lovell NH, Wallace GG, Poole-Warren LA. Conducting polymers for neural interfaces: Challenges in developing an effective long-term implant. Biomaterials 2008;29(24):3393–9.

[122] Balint R, Cassidy NJ, Cartmell SH. Conductive polymers: towards a smart biomaterial for tissue engineering. Acta Biomater 2014;10(6):2341–53.

[123] George PM, Lyckman AW, LaVan DA, Hegde A, Leung Y, Avasare R, et al. Fabrication and biocompatibility of polypyrrole implants suitable for neural prosthetics. Biomaterials 2005;26(17):3511–9.

[124] Simon DT, Carter SA. Electrosynthetically patterned conducting polymer films for investigation of neural signaling. J Chem Phys 2006;124(20):204709.

[125] Xie J, Macewan MR, Willerth SM, Li X, Moran DW, Sakiyama-Elbert SE, et al. Conductive core-sheath nanofibers and their potential application in neural tissue engineering. Adv Funct Mater 2009;19(14):2312–8.

[126] Thompson BC, Richardson RT, Moulton SE, Evans AJ, O'Leary S, Clark GM, et al. Conducting polymers, dual neurotrophins and pulsed electrical stimulation—dramatic effects on neurite outgrowth. J Control Rel Off J Control Rel Soc 2010;141(2):161–7.

[127] Huang J, Hu X, Lu L, Ye Z, Zhang Q, Luo Z. Electrical regulation of Schwann cells using conductive polypyrrole/chitosan polymers. J Biomed Mater Res A 2010;93(1):164–74.

[128] Lakard S, Herlem G, Valles-Villareal N, Michel G, Propper A, Gharbi T, et al. Culture of neural cells on polymers coated surfaces for biosensor applications. Biosens Bioelectron 2005;20(10):1946–54.

[129] Stauffer WR, Cui XT. Polypyrrole doped with 2 peptide sequences from laminin. Biomaterials 2006;27(11):2405–13.

[130] Lee J-W, Serna F, Nickels J, Schmidt CE. Carboxylic acid-functionalized conductive polypyrrole as a bioactive platform for cell adhesion. Biomacromolecules 2006;7(6):1692–5.

[131] Wadhwa R, Lagenaur CF, Cui XT. Electrochemically controlled release of dexamethasone from conducting polymer polypyrrole coated electrode. J Control Release 2006;110(3):531–41.

[132] Nickels JD, Schmidt CE. Surface modification of the conducting polymer, polypyrrole, via affinity peptide. J Biomed Mater Res A 2013;101((5):1464–71.

[133] Gomez N, Schmidt CE. Nerve growth factor-immobilized polypyrrole: Bioactive electrically conducting polymer for enhanced neurite extension. J Biomed Mater Res A 2007;81(1):135–49.

[134] Lee JY, Lee J-W, Schmidt CE. Neuroactive conducting scaffolds: nerve growth factor conjugation on active ester-functionalized polypyrrole. J R Soc Interface 2008;0403. 2008:rsif.

[135] Gomez N, Lee JY, Nickels JD, Schmidt CE. Micropatterned polypyrrole: a combination of electrical and topographical characteristics for the stimulation of cells. Adv Funct Mater 2007;17(10):1645–53.

[136] Zhang L, Stauffer WR, Jane EP, Sammak PJ, Cui XT. Enhanced differentiation of embryonic and neural stem cells to neuronal fates on laminin peptides doped polypyrrole. Macromol Biosci 2010;10(12):1456–64.
[137] Lundin V, Herland A, Berggren M, Jager EW, Teixeira AI. Control of neural stem cell survival by electroactive polymer substrates. PLoS ONE 2011;6(4):e18624.
[138] Saigal R, Cimetta E, Tandon N, Zhou J, Langer R, Young M, et al., editors. Electrical stimulation via a biocompatible conductive polymer directs retinal progenitor cell differentiation. 2013 35th Annual international conference of the IEEE Engineering in Medicine and Biology Society (EMBC); 2013: IEEE.
[139] Stewart E, Kobayashi NR, Higgins MJ, Quigley AF, Jamali S, Moulton SE, et al. Electrical stimulation using conductive polymer polypyrrole promotes differentiation of human neural stem cells: a biocompatible platform for translational neural tissue engineering. Tissue Eng C Methods 2014;21(4):385–93.
[140] Yang S, Jang L, Kim S, Yang J, Yang K, Cho SW, et al. Polypyrrole/alginate hybrid hydrogels: electrically conductive and soft biomaterials for human mesenchymal stem cell culture and potential neural tissue engineering applications. Macromol Biosci 2016;16(11):1653–61.
[141] McDonald JC, Whitesides GM. Poly (dimethylsiloxane) as a material for fabricating microfluidic devices. Acc Chem Res 2002;35(7):491–9.
[142] Pavesi A, Piraino F, Fiore GB, Farino KM, Moretti M, Rasponi M. How to embed three-dimensional flexible electrodes in microfluidic devices for cell culture applications. Lab Chip 2011;11(9):1593–5.
[143] Bani-Yaghoub M, Tremblay R, Voicu R, Mealing G, Monette R, Py C, et al. Neurogenesis and neuronal communication on micropatterned neurochips. Biotechnol Bioeng 2005;92(3):336–45.
[144] Takayama Y, Moriguchi H, Kotani K, Suzuki T, Mabuchi K, Jimbo Y. Network-wide integration of stem cell-derived neurons and mouse cortical neurons using microfabricated co-culture devices. Biosystems 2012;107(1):1–8.
[145] Sun Y, Yong KM, Villa-Diaz LG, Zhang X, Chen W, Philson R, et al. Hippo/YAP-mediated rigidity-dependent motor neuron differentiation of human pluripotent stem cells. Nat Mater 2014;13(6):599–604.
[146] Southam KA, King AE, Blizzard CA, McCormack GH, Dickson TC. Microfluidic primary culture model of the lower motor neuron–neuromuscular junction circuit. J Neurosci Methods 2013;218(2):164–9.
[147] Pedraza E, Brady A-C, Fraker CA, Stabler CL. Synthesis of macroporous poly (dimethylsiloxane) scaffolds for tissue engineering applications. J Biomater Sci Polym Ed 2013;24(9):1041–56.
[148] Barros D, Freitas Amaral I, Paula Pego A. Biomimetic synthetic self-assembled hydrogels for cell transplantation. Curr Top Med Chem 2015;15(13):1209–26.
[149] Koss K, Unsworth L. Neural tissue engineering: bioresponsive nanoscaffolds using engineered self-assembling peptides. Acta Biomater 2016;44:2–15.
[150] Yokoi H, Kinoshita T, Zhang S. Dynamic reassembly of peptide RADA16 nanofiber scaffold. Proc Natl Acad Sci U S A 2005;102(24):8414–9.
[151] Ellis-Behnke RG, Liang Y-X, You S-W, Tay DK, Zhang S, So K-F, et al. Nano neuro knitting: peptide nanofiber scaffold for brain repair and axon regeneration with functional return of vision. Proc Natl Acad Sci U S A 2006;103(13):5054–9.
[152] Gelain F, Bottai D, Vescovi A, Zhang S. Designer self-assembling peptide nanofiber scaffolds for adult mouse neural stem cell 3-dimensional cultures. PLoS ONE 2006;1(1):e119.

[153] Gelain F, Lomander A, Vescovi AL, Zhang S. Systematic studies of a self-assembling peptide nanofiber scaffold with other scaffolds. J Nanosci Nanotechnol 2007;7(2):424–34.

[154] Koutsopoulos S, Zhang S. Long-term three-dimensional neural tissue cultures in functionalized self-assembling peptide hydrogels, matrigel and collagen I. Acta Biomater 2013;9(2):5162–9.

[155] Thonhoff JR, Lou DI, Jordan PM, Zhao X, Wu P. Compatibility of human fetal neural stem cells with hydrogel biomaterials in vitro. Brain Res 2008;1187:42–51.

[156] Liu Y, Ye H, Satkunendrarajah K, Yao GS, Bayon Y, Fehlings MG. A self-assembling peptide reduces glial scarring, attenuates post-traumatic inflammation and promotes neurological recovery following spinal cord injury. Acta Biomater 2013;9(9):8075–88.

[157] Zweckberger K, Ahuja CS, Liu Y, Wang J, Fehlings MG. Self-assembling peptides optimize the post-traumatic milieu and synergistically enhance the effects of neural stem cell therapy after cervical spinal cord injury. Acta Biomater 2016;42:77–89.

[158] Iwasaki M, Wilcox JT, Nishimura Y, Zweckberger K, Suzuki H, Wang J, et al. Synergistic effects of self-assembling peptide and neural stem/progenitor cells to promote tissue repair and forelimb functional recovery in cervical spinal cord injury. Biomaterials 2014;35(9):2617–29.

[159] Zhao X, Yao GS, Liu Y, Wang J, Satkunendrarajah K, Fehlings M. The role of neural precursor cells and self assembling peptides in nerve regeneration. J Otolaryngol-Head Neck Surg 2013;42(1):1.

[160] Zweckberger K, Liu Y, Wang J, Forgione N, Fehlings MG. Synergetic use of neural precursor cells and self-assembling peptides in experimental cervical spinal cord injury. J Vis Exp 2015;(96)e52105-e.

[161] Zhang S, Anderson MA, Ao Y, Khakh BS, Fan J, Deming TJ, et al. Tunable diblock copolypeptide hydrogel depots for local delivery of hydrophobic molecules in healthy and injured central nervous system. Biomaterials 2014;35(6):1989–2000.

[162] Klein R, Blaschke S, Neumaier B, Endepols H, Graf R, Keuters M, et al. The synthetic NCAM mimetic peptide FGL mobilizes neural stem cells in vitro and in vivo. Stem Cell Rev Rep 2014;10(4):539–47.

[163] Huang ML, Smith RA, Trieger GW, Godula K. Glycocalyx remodeling with proteoglycan mimetics promotes neural specification in embryonic stem cells. J Am Chem Soc 2014;136(30):10565–8.

[164] Freese A, Sabel BA, Saltzman WM, During MJ, Langer R. Controlled release of dopamine from a polymeric brain implant: in vitro characterization. Exp Neurol 1989;103(3):234–8.

[165] Saltzrnan WM. Controlled release of dopamine from a polymeric brain implant: in vivo characterization. Ann Neurol 1989;25(4).

[166] Krewson CE, Klarman ML, Saltzman WM. Distribution of nerve growth factor following direct delivery to brain interstitium. Brain Res 1995;680(1):196–206.

[167] Krewson CE, Saltzman WM. Transport and elimination of recombinant human NGF during long-term delivery to the brain. Brain Res 1996;727(1):169–81.

[168] Mahoney MJ, Saltzman WM. Millimeter-scale positioning of a nerve-growth-factor source and biological activity in the brain. Proc Natl Acad Sci 1999;96(8):4536–9.

[169] Saltzman WM, Mak MW, Mahoney MJ, Duenas ET, Cleland JL. Intracranial delivery of recombinant nerve growth factor: release kinetics and protein distribution for three delivery systems. Pharm Res 1999;16(2):232–40.

[170] A Stout D. Recent advancements in carbon nanofiber and carbon nanotube applications in drug delivery and tissue engineering. Curr Pharm Des 2015;21(15):2037–44.
[171] Bokara KK, Kim JY, Lee YI, Yun K, Webster TJ, Lee JE. Biocompatability of carbon nanotubes with stem cells to treat CNS injuries. Anatomy Cell Biol 2013;46(2):85–92.
[172] Kim JA, Jang EY, Kang TJ, Yoon S, Ovalle-Robles R, Rhee WJ, et al. Regulation of morphogenesis and neural differentiation of human mesenchymal stem cells using carbon nanotube sheets. Integr Biol 2012;4(6):587–94.
[173] Huang YJ, HC W, Tai NH, Wang TW. Carbon nanotube rope with electrical stimulation promotes the differentiation and maturity of neural stem cells. Small 2012;8(18):2869–77.
[174] Kabiri M, Soleimani M, Shabani I, Futrega K, Ghaemi N, Ahvaz HH, et al. Neural differentiation of mouse embryonic stem cells on conductive nanofiber scaffolds. Biotechnol Lett 2012;34(7):1357–65.
[175] Sridharan I, Kim T, Wang R. Adapting collagen/CNT matrix in directing hESC differentiation. Biochem Biophys Res Commun 2009;381(4):508–12.
[176] Kim T, Sridharan I, Zhu B, Orgel J, Wang R. Effect of CNT on collagen fiber structure, stiffness assembly kinetics and stem cell differentiation. Mater Sci Eng C 2015;49:281–9.
[177] Hackelbusch S, Rossow T, Steinhilber D, Weitz DA, Seiffert S. Hybrid microgels with thermo-tunable elasticity for controllable cell confinement. Adv Healthcare Mater 2015;4(12):1841–8.
[178] Li Z, Fan Z, Xu Y, Lo W, Wang X, Niu H, et al. pH-sensitive and thermosensitive hydrogels as stem-cell carriers for cardiac therapy. ACS Appl Mater Interfaces 2016;8(17):10752–60.
[179] Highley CB, Kim M, Lee D, Burdick JA. Near-infrared light triggered release of molecules from supramolecular hydrogel-nanorod composites. Nanomedicine (Lond) 2016;11(12):1579–90.
[180] Xiao Y-Y, Gong X-L, Kang Y, Jiang Z-C, Zhang S, Li B-J. Light-, pH-and thermal-responsive hydrogels with the triple-shape memory effect. Chem Commun 2016;52(70):10609–12.
[181] Kim H, Park H, Lee JW, Lee KY. Magnetic field-responsive release of transforming growth factor beta 1 from heparin-modified alginate ferrogels. Carbohydr Polym 2016;151:467–73.
[182] Weng L, Xie J. Smart electrospun nanofibers for controlled drug release: Recent advances and new perspectives. Curr Pharm Des 2015;21(15):1944–59.
[183] Boas M, Gradys A, Vasilyev G, Burman M, Zussman E. Electrospinning polyelectrolyte complexes: pH-responsive fibers. Soft Matter 2015;11(9):1739–47.
[184] Jiang J, Xie J, Ma B, Bartlett DE, Xu A, Wang C-H. Mussel-inspired protein-mediated surface functionalization of electrospun nanofibers for pH-responsive drug delivery. Acta Biomater 2014;10(3):1324–32.
[185] Yuan H, Li B, Liang K, Lou X, Zhang Y. Regulating drug release from pH-and temperature-responsive electrospun CTS-g-PNIPAAm/poly (ethylene oxide) hydrogel nanofibers. Biomed Mater 2014;9(5):055001.
[186] Kim Y-J, Ebara M, Aoyagi T. Temperature-responsive electrospun nanofibers for 'on–off'switchable release of dextran. Sci Technol Adv.Mater 2016;13(6):064203.

CHAPTER

Drug delivery systems for engineering neural tissue

7

1 INTRODUCTION

Different types of pharmacological agents such as growth factors and small molecules, can manipulate the biological functions of the cells present in the body [1–3]. Often, strategies for tissue engineering take advantage of these properties to induce regeneration post-injury or to control stem cell behavior. These molecules often have short half-lives as they become degraded by enzymes or the immune system clears them from the body. The field of drug delivery represents a broad field as this term refers to any attempt or approach to extend and/or target the activity of a pharmacological agent intended to have a desired biological effect in the body. While developing extended release, formulations of pharmaceutical agents from tablets and capsules can be considered as part of drug delivery. This chapter will focus on the applications of drug delivery in tissue engineering. As discussed in detail in Chapters 5 and 6, many different types of natural and synthetic polymers have been used as scaffolds for cell delivery often in combination with drug delivery of appropriate agents. Thus, scientists have developed a variety of drug delivery systems based on polymer scaffolds as a way to provide extended and controlled release of these different types of molecules [4].

Drug delivery systems can also provide an effective way of delivering biological agents to targeted regions of the body in a controlled fashion. Design considerations for such systems include the properties of the molecules to be delivered, such as its charge, hydrophobicity, molecular weight, and chemical structure, the desired delivery site in the body, effective concentration needed for the desired effect, and the necessary time course for release. Many tissue engineering strategies encapsulate biomolecules into polymer scaffolds. Accordingly, the bioactivity of the molecule must be confirmed post encapsulation to ensure that the incorporation process did not negatively impact the molecule being delivered. Additional considerations for clinical applications include the stability of the drug over time, which will correspond to the shelf life of the product, and how such a system will be delivered in vivo as often surgeons prefer injectable scaffolds. This section will also address the benefits and limitation of these different types of drug delivery systems in the context of engineering neural tissue from stem cells. Often the molecules necessary for preserving existing tissue in the damaged nervous system require a different rate and concentration of release compared to the factors necessary for controlling stem cell behaviors. Thus, an ideal scaffold may incorporate a variety of drug delivery systems for releasing multiple factors.

Engineering Neural Tissue from Stem Cells. http://dx.doi.org/10.1016/B978-0-12-811385-1.00007-8

Growth factors, along with a subset of neural related proteins known as neurotrophins, are highly desirable molecules for treating diseases and disorders of the nervous system as discussed in Chapter 4. Accordingly, growth factors are one of the most common targets incorporated into drug delivery systems developed for applications in engineering neural tissue [5–7]. However, despite their positive benefits, proteins are sensitive to extreme changes in temperature, pH, and salt concentrations. In addition, small molecules can modulate the behavior of the cells present in the diseased or damaged nervous system as well as stem cells that have been transplanted into the nervous system. These molecules tend to be easier to deliver using diffusion-based delivery systems due to their size and their inability to be denatured unlike proteins. Other popular molecules used to treat disorders of the nervous system include anti-inflammatories, which modulate the body's response to injury by reducing pain and swelling [8,9]. Other biomolecules work by reducing the inhibitory effects that result from damage to central nervous system. A class of enzymes known as chondroitinases degrade inhibitory proteoglycans expressed after traumatic injuries that actively prevent regeneration attempts by the central nervous system [10–12]. However, maintaining the activity of these enzymes remains challenging, making it difficult to deliver them over extended time courses. Thus, their stability and function must be considered when determining an appropriate drug delivery strategy. Other molecules that can reduce the inhibitory environment post-trauma include antibodies against different components released by damaged oligodendrocytes, such as Nogo, myelin-associated glycoproteins (MAGs), and oligodendrocyte-myelin glycoproteins (Omgp) [13]. These components bind a common receptor known as Nogo, which transmits signals through RhoA. Antibodies that bind to these signals promote enhanced recovery post-injury, suggesting that this pathway can determine how permissive an environment is post-injury. Finally, several drugs are currently used to treat the symptoms of different neurological diseases as opposed to representing long-term cures as discussed in Chapter 1. Thus, they are not typically used for tissue engineering applications.

This section will cover the major categories of drug delivery systems that can be incorporated into biomaterial scaffolds to enhance their properties. Each of these delivery methods will be discussed in terms of what biomolecules they are capable of delivering, their advantages, and their drawbacks and limitations. As detailed examples of these systems have been discussed in depth during Chapters 5 and 6, this section will focus mainly on the rationale for using each of these types of delivery systems along with their characteristic features. The first section will address systems that rely on the intrinsic characteristics of the material to regulate the release of drugs through diffusion. These systems can be fabricated in a wide variety of formulations, including hydrogels, microparticles, nanoparticles, microfibers, and nanofibers, and this chapter will discuss their properties. It then moves onto affinity-based drug delivery systems, which use known interactions between a target biomaterial scaffold and a target molecule to generate sustained release of that molecule over time. Many affinity-based drug delivery systems incorporate the polysaccharide heparin into a biomaterial to enhance its ability to bind a variety of growth factors, including the members of the neurotrophin family. This chapter will also discuss the various chemical schemes that can be used to physically tether bioactive factors to biomaterial scaffolds as way of presenting these cues as well as discussing systems that rely on electrical stimulation to release bioactive factors. Finally, the use of optically responsive materials for drug delivery will be evaluated in the context of neural tissue engineering.

2 DIFFUSION-BASED DRUG DELIVERY SYSTEMS

In accordance with Fick's law, molecules move from areas with a higher concentration to areas where there is a lower concentration as the system attempts to reach equilibrium. This principle serves as the basis for the design of diffusion-based drug delivery systems. Accordingly, the most commonly used drug delivery systems incorporate bioactive molecules into a material and then they rely on the material properties to modulate the release rate of this target molecule by diffusion. Using different polymer materials modulates the rate of diffusion, which in turn regulates how fast these molecules diffuse. These diffusion-based release systems come in many forms as detailed in the following sections. One of the major limitations of such systems is that their drug release rate relies on the inherent material properties of the scaffold. Many different formulations of such systems, including hydrogels, particles, and fibers, have been produced as a way of addressing this issue associated with release rates. Changing the structure of the scaffold affects the rate of release. Another way of addressing this issue involves combining multiple forms of diffusion-based drug delivery systems. For example, hydrogels can be loaded with drug-releasing nano- and microparticles. Or alternatively, fibers can be incorporated into hydrogels. Both proposed systems could have their constituent components altered to have different rates of release. In addition, structural differences also create variance in drug delivery between the two different types of scaffold structures.

2.1 HYDROGELS

Hydrogels serve as one of the most commonly used polymer scaffold formulations for delivery of stem cells and drugs [14–19]. Hydrogels consist of hydrophilic polymer chains cross-linked together that swell when exposed to an aqueous environment, allowing them to contain a high level of water content (up to 90% of their total mass). This section focuses on the advantages and disadvantages of using hydrogels that rely on their physical properties to generate extended release of biological molecules over time as these scaffolds have been discussed in detail in the previous two chapters. These factors influencing the rate of drug release include the biomaterial used for fabrication, the degree of polymer cross-linking, and the concentration of the target molecule to be delivered. Injectable hydrogels can be formulated, which is a desirable property for preclinical and clinical applications [20]. With their inherent properties regulating diffusion, one of the major advantages of using hydrogels to deliver drugs is that the fabrication process can be straightforward for a number of polymers. It only requires incorporation of these molecules during the fabrication process, which for most natural polymers and some synthetic polymers require mild conditions that do not negatively impact bioactivity. For example, the Tuszynski group incorporated a variety of growth factors into fibrin scaffolds to enhance the survival and integration of stem cells post transplantation into an in vivo preclinical model of spinal cord injury [21–23]. One of the major disadvantages of relying on inherent material properties is that these properties, along with the degradation rate of the material, control the release rate of the target molecule. Often this time course does not correspond with the most desirable time courses and release rates for a specific tissue engineering application. However, these limitations can be addressed by using some of the other scaffold formulations detailed below in combination with hydrogels.

2.2 MICROPARTICLES

Microparticles range in size from 0.1 to 100 μm. This range of sizes gives them several desirable properties for applications in drug delivery. These particles can be fabricated using a variety of techniques that usually involve making an emulsion to incorporate bioactive molecules into them [24]. Other methods use microfluidic devices to generate these particles [25]. Microfluidic methods, which will be discussed in later sections, use tools to manipulate low volumes of fluids and molecules in the range of microliters for a wide variety of applications. Microspheres fabricated from poly(lactic-*co*-glycolic acid) (PLGA) using a batch emulsion process can release bioactive versions of nerve growth factor (NGF) and glial-derived neurotrophic factor (GDNF) [26,27]. My group has also demonstrated that poly(caprolactone) (PCL) microspheres could release bioactive GDNF in a controlled fashion over a month [28]. Fig. 1 shows a representative image of such microspheres. Their size also means they

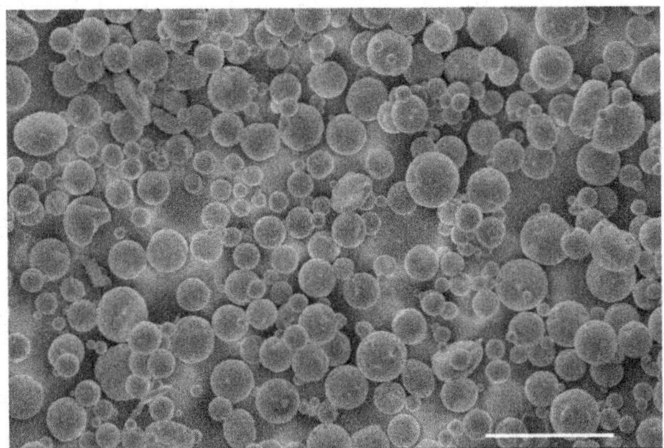

FIG. 1

Scanning electron microscopy images of microspheres fabricated from poly(caprolactone) that contain biologically active glial-derived neurotrophic factor. Scale bar is 500 μm.

Image taken by Andrew Agbay—graduate student in Willerth lab using the Advanced Microscopy Facility at the University of Victoria.

can be injected into the desired site to deliver the target molecule. For example, in Parkinson's disease, the dopaminergic neurons located in the *substantia nigra* region of the brain die off. Thus, when microspheres that release GDNF were injected into this region in a mouse model of the disease, they reversed some of these effects by promoting cellular survival [29]. Similar approaches injected GDNF-releasing microspheres into the retina were able to promote photoreceptor survival in a mouse model of retinal degeneration [30].

Other benefits of using microspheres include being able to tune the concentration of drug being delivered by changing the number of microspheres being injected or implanted. Specific factors that control the target release rate include the size of particles and the biomaterial used to fabricate these particles. One concern regarding this type of scaffold is that the use of solvents to fabricate these microparticles can have a negative impact on growth factor activity, which is why it is important to confirm bioactivity after encapsulation. Another issue involves ensuring that the proper amount of drug necessary to achieve a biological effect has been encapsulated inside of these particles. These particles are also too big to cross the blood-brain barrier that separates the central nervous system from the rest of the body. Navigating the blood-brain barrier is a major concern when engineering neural tissue. The methods used to fabricate microparticles can often produce a wide range of particle sizes. However, sieves can be used to ensure a more homogenous size distribution of particles.

As discussed in Chapter 6, microspheres can be combined with hydrogels to provide multiple ways of delivering different therapeutic drugs with varying rates of release. Also, microspheres containing different types of drugs can be incorporated in such systems as a way of generating combinatorial therapies for treating the diseases and disorders of the central nervous system. For example, one set of microparticles could deliver anti-inflammatory drugs to reduce the harsh microenvironment that occurs post-traumatic injury while another set of particles could deliver regeneration promoting growth factors. Overall, microspheres serve as a useful tool for engineering neural tissue—both on their own and when used in combination with other types of scaffolds.

2.3 NANOPARTICLES

Nanoparticles, as their name suggests, consists of particles with sizes in the nanoscale range [31]. Their size makes it easy to deliver these particles to their target location using injection and such particles have been fabricated from a variety of substances ranging from polymers to metals. These particles deliver a wide variety of cargos, including proteins, DNA, RNA, and other molecules [32,33] with the properties of the material used to fabricate the particles along with the size of the resulting particles influencing the release rate [34]. They can also be synthesized into different shapes, and various cellular populations can phagocytose these different shaped nanoparticles at variable rates [35]. Nanoparticles can also be functionalized with antibodies and other chemical moieties such that these particles target specific cellular populations, including those found in the central nervous system [36]. For example, nanoparticles can be designed to target receptors associated with neurotransmitters expressed by a subset of neural cells, like neurons [37]. While many studies have focused on the development of nanoparticles that target neurons, few studies have looked at how to target glial cell populations [36]. Developing nanoparticles that can target glial cells remains an important challenge to be addressed and it represents a significant opportunity for future work. Such multifunctional nanoparticles can also direct the differentiation of stem cells into the desired neural phenotypes through targeted delivery of the appropriate cargo [38]. A significant amount of research has been done on the use of nanoparticles to deliver anticancer medicine to treat glioblastoma, suggesting these drug delivery systems could be used for other applications in engineering neural tissue [39]. One of the major advantages based on the size of nanoparticles is that they can be transported across the blood-brain barrier, which is difficult to cross for many types of drugs [40]. These nanoparticles can be modified so that their surface chemistry facilitates transport across this structure, enabling them to target the cells of the central nervous system. In addition to being used as a way to control cell behavior, nanoparticles are often used to image structures present in the nervous system to diagnose disease and damage [41]. In terms of limitations, nanoparticles can induce toxic effects in the central nervous system, which is an undesirable feature [42]. Thus, an ideal nanoparticle would be able to simultaneously deliver its biological cargo to a target population of neural cells while minimizing the potential for neurotoxicity. These drug delivery systems have a significant potential as a way for treating nervous system diseases and disorders.

2.4 MICROFIBERS

Microfibers can also serve as drug reservoirs for applications in neural tissue engineering. Similar to microspheres, their size ranges in the microscale. Fig. 2 shows a representative image of a microfiber scaffold. These thin fibers can be fabricated

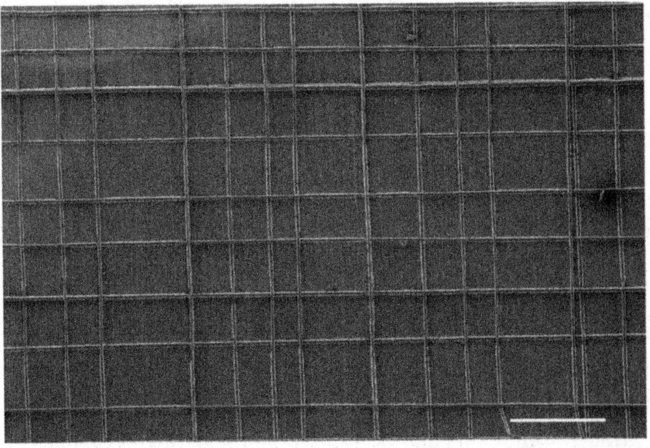

FIG. 2

Scanning electron image of microfibers produced using melt electrospinning. Scale bar is 500 μm.

Image taken by Dr. Nima Mohtaram when he was a graduate student in Willerth lab using the Advanced Microscopy Facility at the University of Victoria.

using a variety of techniques, including melt electrospinning as detailed in Chapter 6. They possess many advantages as a drug delivery system, including the ability to present novel topographical and mechanical cues in addition to the chemical ones provided by controlled drug delivery. However, the conditions for fabricating microfibers can be quite harsh depending on the fabrication method and unique approaches can be used to incorporate bioactive factors into such microfiber scaffolds. For example, a 2013 study impregnated PCL microfibers with silicon nanoparticles, enabling diffusion-based drug release for applications in ophthalmology [43]. They were able to tune the release of silicon from these scaffolds by varying the concentration of nanoparticles used. These scaffolds were compatible as substrates from culturing epithelial cells isolated from the lens of the eye, suggesting their potential applications in neural tissue engineering. The goal would be to encapsulate the desired biomolecules into the silicon before addition to the PCL microfibers. Intriguingly, a pair of studies from the Mathiowitz group at Brown University found that incorporation of the small molecule dexamethasone and proteins with different molecular weights (insulin, lysozyme, and bovine serum albumin) into microfibers scaffolds fabricated from PLGA and poly(L-lactic acid) (PLLA) resulted in diffusion-based controlled release for up to 63 days [44,45]. They also examined how the addition of such proteins influenced the mechanical properties of the microfibers. In their study focused on delivering proteins, they used a unique cryogenic emulsion technique to ensure the successful incorporation of proteins into these microfibers.

Other approaches to making drug-releasing microfibers rely on protein crystallization to retain bioactivity [46]. Despite their immense potential, there have been relatively few studies using drug-releasing microfibers for neural tissue engineering applications in general. My group, detailed in Chapter 6, functionalized PCL microfibers with retinoic acid-releasing nanofibers as way to enhance their bioactivity [47]. However, most microfiber fabrication techniques use harsh solvents or high temperatures making delivery of neurotrophic proteins difficult. Novel methods are required to incorporate successfully protein-based drug release to such scaffolds. However, once functionalized, these scaffolds could be applied as patches to serve as an artificial dura and their topographical cues can also be used to direct stem cell behavior as well in addition to producing higher-level architecture that mimics the extracellular matrix [48]. Thus, further exploration and characterization of drug-releasing microfibers remains an intriguing area for future research to explore. Accordingly, one promising technique that will be explored in more detail in the next section is the use of microfluidic devices to fabricate fibers of with microscale and nanoscale topography, which avoids some of the issues associated with other techniques for manufacturing microfibers that result in loss of bioactivity [49]. Fig. 3 shows an overview of microfluidic spinning as a fiber fabrication technique.

2.5 NANOFIBERS

In comparison with microfibers, a significant body of work detailing how to fabricate drug-releasing nanofibers exists, including the use of such scaffolds for neural tissue applications as detailed in Chapter 6 and in review articles [50–55]. These fibers are

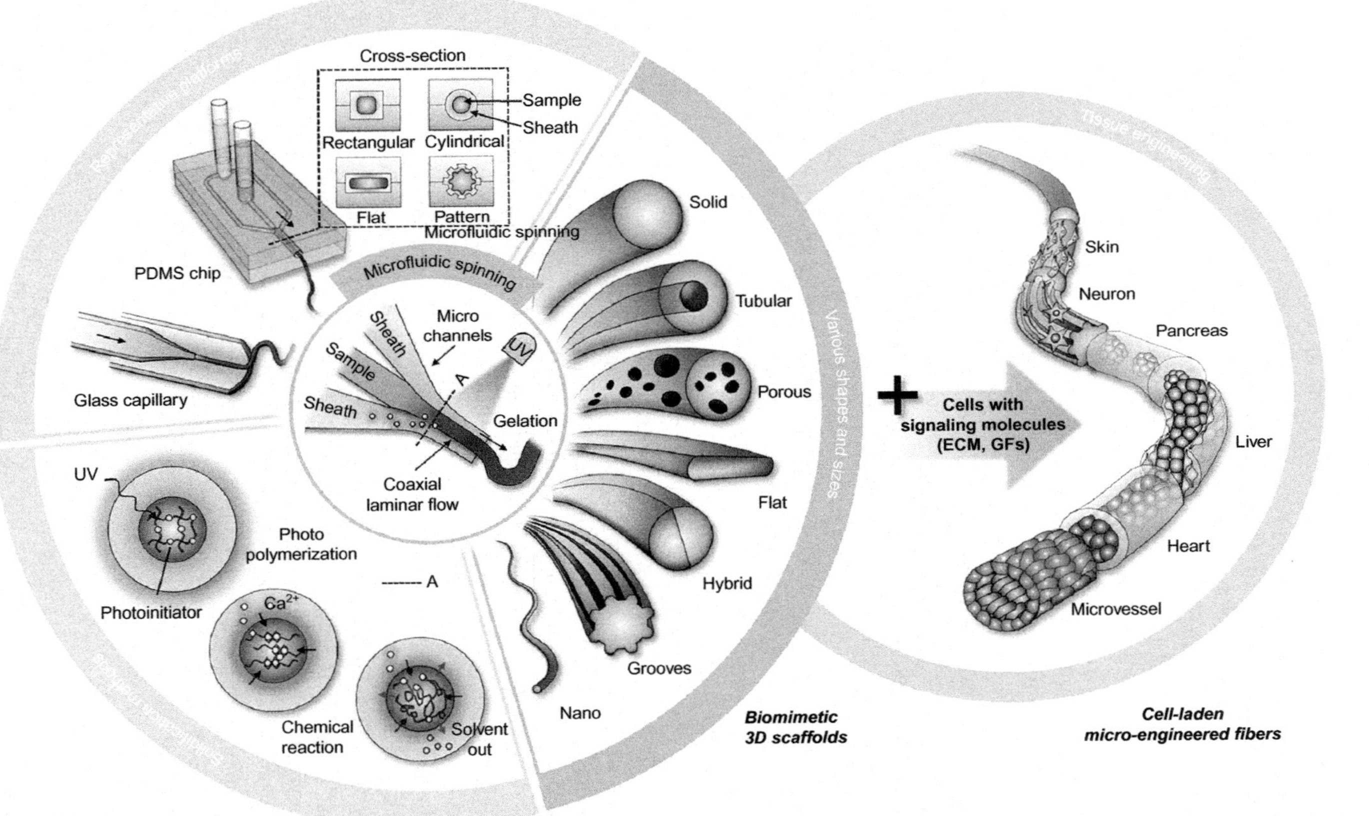

FIG. 3

Overview of the various microfluidic spinning methods used to fabricate fibers, along with the fibers' tissue engineering applications. (Left) Microfluidic platforms, such as pulled glass micropipettes or PDMS-based chips, can be used to prepare a variety of patterned channel shapes. The fibers can be created either by photopolymerizing a sample fluid or through chemical reactions between the sample and sheath fluids undergoing coaxial laminar flow in a microchannel. Depending on the platform and materials, fibers with a variety of shapes and sizes may be produced as biomimetic 3D scaffolds. (Right) For the tissue engineering or organ regeneration applications, specific cells with suitable biochemical and physicochemical factors could be incorporated into engineered fibers and may then be formed into the desired tissue or artificial organ.

Reprinted with permission from Jun Y, Kang E, Chae S, Lee S-H. Microfluidic spinning of micro-and nano-scale fibers for tissue engineering. Lab Chip 2014;14(13):2145–60.

defined by their nanoscale size, which enables the production of scaffolds possessing high surface area and small pore sizes. Such scaffolds tend to have a high level of porosity, enabling cells to infiltrate them. The material selection, the properties of the molecule being incorporated and the fiber size, all modulate the diffusion-based drug-release rate from nanofibers, and nanofibers can produce release time courses of weeks to months. Another benefit of these nanofiber scaffolds is that they can be fabricated in scaffolds with different topographies that can also influence stem cell behavior [56]. Like microfibers, these scaffolds can serve as patches to be used as artificial duras in vivo to repair injuries after trauma or as substrates for stem cell culture and differentiation of stem cells into neural phenotypes. Different methods for encapsulating bioactive agents into nanofibers exists, including the use of emulsions and the use of coaxial electrospinning. In coaxial electrospinning, two sets of solvents are used to fabricate the core and shell regions of the nanofibers, respectively. The use of different solvents and polymers mean that these regions possess different properties. Often the core region is fabricated with solvents that are less likely to affect the bioactivity of the molecule being delivered from the fibers. In recent years, microfluidic spinning as a method for producing such nanofibers has increased in popularity and a recent review compares this method with electrospinning [57]. For example, solution electrospinning often produces fibers that are highly variable in diameter, as fiber size depends on the scaffold material and the target molecule incorporated. My research group has observed that the molecular weight and charge of the molecule being incorporated into the nanofibers during the electrospinning process have a significant effect on the fiber diameter [58,59]. Also, the nanofiber production process often requires the use of harsh solvents that can negatively impact the bioactivity of incorporated proteins. These effects can be minimized by using stabilizing agents or by using coaxial electrospinning to produce nanofibers with defined core and shell regions. Alternatively, microfluidic-based methods can also potentially address these issues when used to fabricate drug-releasing nanofibers as well. Such methods often use different chemical reactions or photopolymerization to fabricate drug-releasing fibers.

3 AFFINITY-BASED DRUG DELIVERY SYSTEMS

Affinity-based drug delivery system uses known interactions between molecules to generate controlled release of bioactive molecules as detailed in recent reviews from the Shoichet lab, which has made significant contributions in terms of developing such drug delivery systems [60–62]. These interactions can be identified based on biological interactions found in nature, interactions observed experimentally in laboratories, or using high throughput screening. Thus, these systems offer more flexibility in comparison to diffusion-based drug delivery systems in terms of tailoring release rates of biomolecules. In addition, a significant amount of work has been investigating how to model these systems in silicio and these mathematical models provide insight to how such systems can be tailored to generate the desired release profile for

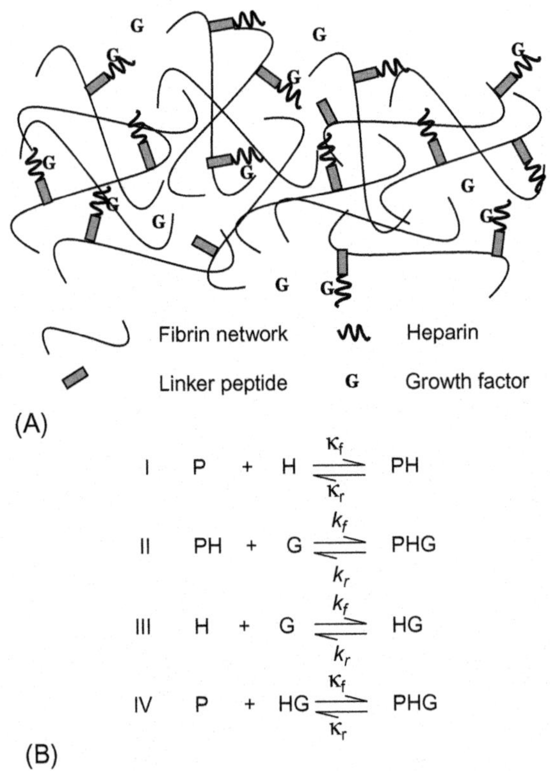

FIG. 4

(A) Diagram showing the components of the heparin-binding delivery system. α2PI1–7–ATIII121–134 peptide is cross-linked into the fibrin gel via the transglutaminase activity of Factor XIIIa; heparin can bind to the peptide by electrostatic interactions. NT-3 can bind to the bound heparin, creating a gel-bound ternary complex that is not diffusible. NT-3 can also exist in the diffusible state, alone, or in a complex with free heparin. (B) Four chemical equations defining association of components in the HBDS. Forward and reverse rates for each association are assumed to be independent of the previous bound state of each component. H denotes heparin, P denotes peptide, G denotes growth factor, κf and κf denote the forward and reverse rates for the heparin/peptide interaction, respectively, and kf and kr denote the forward and reverse rates for heparin/growth factor interaction, respectively.

Reprinted with permission from Taylor SJ, McDonald JW, 3rd, Sakiyama-Elbert SE. Controlled release of neurotrophin-3 from fibrin gels for spinal cord injury. J Control Release 2004;98(2):281–94.

a target molecule [63–67]. As an example of such a model, Fig. 4 shows the composition of a heparin-based delivery system used to deliver growth factors from fibrin scaffolds and the set of equations used to determine the retention of growth factors inside of such a system at equilibrium taken from a 2004 paper from the Sakiyama-Elbert group [63]. Section 3.1 will discuss how heparin can be used as a part of such delivery systems to generate controlled release of growth factors and Section 3.2 will then detail how other types of affinity-based drug delivery systems work and their applications. Overall, the use of affinity to generate controlled release provides a versatile and interesting strategy for drug delivery.

3.1 HEPARIN-BASED DRUG DELIVERY SYSTEMS

The polysaccharide heparin plays important roles in the extracellular matrix where it modulates cell behavior through a variety of mechanisms as detailed in Chapter 4 and elsewhere [68]. One of these mechanisms is the ability of heparin to bind a number of different growth factors with varying affinities, including neurotrophins [69]. Pioneering work into heparin-based delivery systems was conducted initially in the lab of Dr. Jeffrey Hubbell when his research group showed that functionalizing fibrin scaffolds with a heparin-binding peptide could produce controlled release of growth factors when heparin was present in their novel engineered tissues (Fig. 4) [70–72]. They first demonstrated that such a system could successfully deliver biologically active basic growth factor (bFGF) and then that it could deliver neurotrophins, like NGF, that possess lower affinities for heparin. These systems successfully delivered controlled release of NGF in vivo for repairing peripheral nerve injury in rats [73] and associated work clarified how modifying the amount of heparin used in this drug delivery system influenced growth factor release [66]. More work from the Sakiyama-Elbert group used such heparin-based delivery systems as way to deliver neurotrophin-3 (NT-3) for repair of spinal cord injury [63,74,75] and GDNF for peripheral nerve repair [76,77]. My work during graduate school under Dr. Sakiyama-Elbert's guidance adapted these systems to deliver multiple growth factors as a way of engineering neural tissue from mouse embryonic stem cells [78]. Later work evaluated the ability of such engineered tissues to promote regeneration in a preclinical model of spinal cord injury [79,80]. Overall, this large body of work indicates both the feasibility and potential of these heparin-based systems for engineering neural tissue from pluripotent stem cells.

Other work from the Sakiyama-Elbert group used a phage display to identify peptide domains with varying affinities for heparin [64]. These domains were then used as part of the drug delivery system to modulate further growth factor release rates. This work demonstrates how high throughput screening can be performed to identify affinity interactions. In 2013, the Hubbell lab also identified heparin-binding domains present in the protein fibrinogen and they adapted these domains for use in such drug delivery systems [81]. Finally, other work has demonstrated that such delivery systems can be incorporated into scaffolds consisting of chitosan and alginate as a way to create controlled release of bFGF [82]. Future work could adapt heparin-based drug delivery systems for other applications in repairing the damaged nervous system, including delivering stem cells for the treatment of neurodegenerative disorders.

3.2 OTHER TYPES OF AFFINITY-BASED DRUG DELIVERY SYSTEMS

While heparin remains a popular component of affinity-based delivery systems, other work has explored the use of other interactions to generate controlled drug release for applications in neural tissue engineering. These approaches can be broadly classified into (1) interactions found in nature and (2) interactions based on the material properties. For example, a set of studies demonstrated how functionalizing different proteins (bFGF), ciliary neurotrophic factor (CNTF), brain-derived neurotrophic factor (BDNF), and NT-3 with a collagen-binding domains resulted in successful affinity-based delivery from collagen scaffolds [83–86]. The collagen scaffolds delivering bFGF and CNTF were used to treat a sciatic nerve injury in rats while scaffolds delivering BDNF and NT-3 were evaluated in preclinical models of spinal cord injury. The Shoichet group has also investigated the use of alternative affinity interactions for generating controlled growth factor release. They modified their composite hyaluronic acid/methylcellulose (HAMC) scaffolds to contain peptides with peptides that bound Src protein domains with low and high affinity (Fig. 5) [87]. They then characterized the ability of these scaffolds to release recombinant FGF functionalized with an SH3 domain taken from a Src protein, finding that such a delivery system could extend release up to 10 days compared to 48 hours for unmodified HAMC hydrogels. They used a similar delivery method for successful extended delivery of the enzyme chondroitinase ABC from the same type of hydrogels [88]. This work is interesting as delivery of bioactive chondroitinase ABC is very challenging due to its inherent instability and difficulty in maintaining its enzymatic activity. Finally, recent work from their lab has demonstrated that this system-generated controlled release of a fusion form of insulin-like growth factor-1 (IGF-1) from HAMC scaffolds and the controlled IGF-1 release promoted an increase of viability of retinal pigment

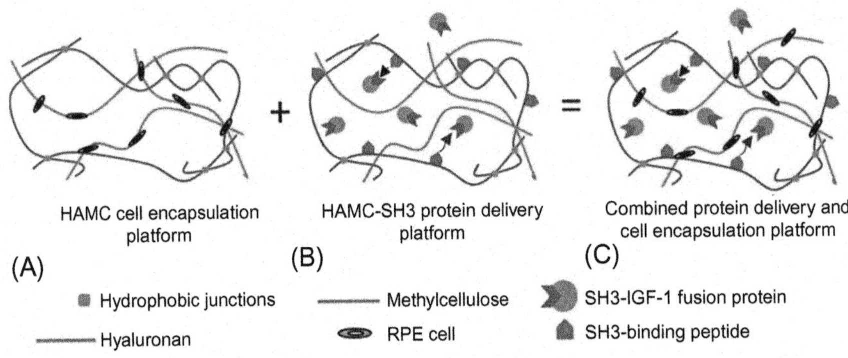

FIG. 5

Schematic of hydrogel designed for combined tunable affinity release of protein therapeutics and cell encapsulation. (A) Hyaluronan and methylcellulose hydrogel (HAMC) known to increase cell viability. (B) Modified HAMC with SH3-binding peptide for the affinity release of SH3–IGF-1 fusion protein. (C) Modified HAMC hydrogel with encapsulated cells.

Reprinted with permission from Parker J, Mitrousis N, Shoichet MS. Hydrogel for simultaneous tunable growth factor delivery and enhanced viability of encapsulated cells in vitro. Biomacromolecules 2016;17(2):476–84.

epithelium cells derived from human embryonic stem cells [89]. This set of studies provides relevant examples of how naturally occurring affinity interactions can be translated successfully into drug delivery systems that generate sustained release of bioactive factors.

Recently, a pair of studies have looked at the use of nanoparticles to generate the necessary electrostatic interactions for successful growth factor delivery. The Stabenfelt group at Arizona State University incorporated nanoparticles into fibrin scaffolds as a way to generate different release rates of the protein stromal cell-derived factor 1α (SDF-1α) [90]. In this study, they found that the fibrin modulated the release of SDF-1α from PLGA nanoparticles and the release rate could be altered by increasing the density of the fibrin scaffold. Thus, the release rate could be tuned through the fabrication parameters of the PLGA microspheres or through changing the properties of the fibrin hydrogel. Finally, the Shoichet group showed in a recent ground breaking study that it was possible to create a encapsulation free drug delivery system based on affinity interactions between nanoparticles embedded in a hydrogel and growth factors (Fig. 6) [91]. As discussed in this chapter and previous chapters, one of the most common ways of generating diffusion-based release of factors requires them to be encapsulated into a polymer. Here the electrostatic interactions between the nanoparticles and the growth factors generate the controlled release as the factors adsorb to the surfaces of these nanoparticles. The strength of

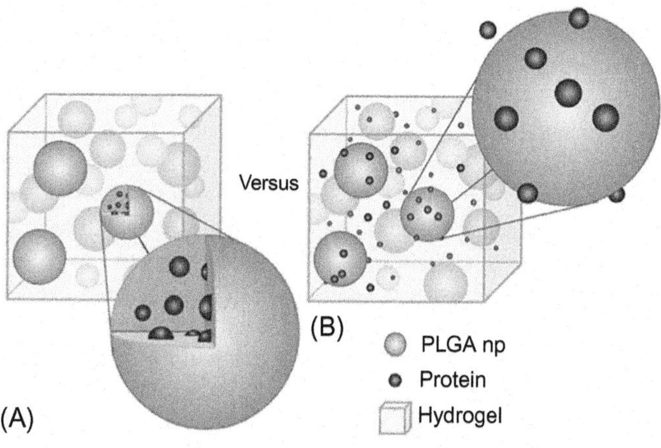

FIG. 6

Two different PLGA nanoparticle systems are compared for controlled protein release. (A) Protein encapsulated in PLGA nanoparticles dispersed in a hydrogel. (B) Protein and blank PLGA nanoparticles dispersed in a hydrogel. For the latter, protein adsorbs to the PLGA nanoparticles but is not encapsulated within them.

Taken from Pakulska MM, Donaghue IE, Obermeyer JM, Tuladhar A, McLaughlin CK, Shendruk TN, et al. Encapsulation-free controlled release: electrostatic adsorption eliminates the need for protein encapsulation in PLGA nanoparticles. Sci Adv 2016;2(5):e1600519; This figure is reprinted under the terms of the Creative Commons Attribution—NonCommercial license.

these interactions determines what type of release profile results. In particular, they found that SDF-1α, NT-3, and BDNF all have similar release from their system while erythropoietin could not be released in a controlled fashion. The release profile could be further tuned varying nanoparticle size and concentration, as well as by changing the pH. Overall, these two studies provide interesting starting points for future work into how nanoparticles can be used as essential components of affinity-based delivery systems. The benefit of using this system is that it does not require a known interaction to be identified in nature, although as detailed above, this system is not capable of generating controlled release of all types of biomolecules.

4 IMMOBILIZATION-BASED DRUG DELIVERY SYSTEMS

Another approach to functionalizing scaffolds with biological molecules is to directly tether the molecules of interest to a biomaterial scaffold. The resulting scaffolds can be used to culture undifferentiated pluripotent stem cells or present factors that drive stem cells to differentiate into lineages relevant for neural tissue engineering applications. Several examples of this approach were detailed in Chapter 6 and this section will detail select relevant examples of how this strategy has been applied to engineering neural tissue. The idea of tethering a biomolecule to a scaffold is straightforward in terms of simplicity as a strategy for drug delivery. Covalently bonding the target molecule to a material ensures that the drug remains in the scaffold system instead of diffusing out. Thus, molecules such as heparin and collagen can be tethered to surfaces to provide suitable substrates for pluripotent stem cell culture [92,93].

The process of tethering can deliver contrasting cues to ensure proper patterning of neurons as done in a recent study from the Leipzig group where substrates presented varying concentrations of NGF and semaphorin-3A to control neurite outgrowth [94]. The tethering process can also produce concentration gradients of proteins under the appropriate conditions [95]. Such gradients can be used to direct cell migration or neurite outgrowth. The immobilization process can also enhance growth factor stability in certain situations. A recent study demonstrated that GDNF covalently tethered to PCL scaffolds remained attached for 120 days [96]. The presented GDNF activated the Erk signaling cascade in neural progenitors, indicating the retention of bioactivity. However, despite all these advantages, such drug delivery systems come with a few caveats. First, the tethering process may interfere with the ability of the cell to process the biomolecules as some cell signaling events require internalization to proceed. In addition, if the material begins to break down, the factors will be released from the scaffold which may or may not be desirable, depending on the desired concentration necessary for achieving a certain biological effect. Finally, appropriate chemistries must be used during the tethering process to ensure that the bioactivity of the molecule is retained. Overall, immobilization-based drug delivery has significant applications for culturing pluripotent stem cells as well as for neural tissue engineering.

5 ELECTRICALLY CONTROLLED DRUG DELIVERY SYSTEMS

As the nervous system functions by sending and receiving electric signals that in turn cause the release of chemical signals, it makes sense to develop drug delivery methods that respond to such electric signals by releasing chemical cues. Accordingly, many of the drug delivery systems that rely on such signals are incorporated into electrodes, the conductive devices used to send electrical signals into nonmetallic materials including neural tissue. Electrodes are often used in the brain to record the activity of nerve cells and thus many of the studies have incorporated such electrically controlled drug delivery systems into electrodes. For example, work from the Cui group showed that electrical simulation could be used to release the steroid dexamethasone from polypyrrole electrodes [97]. The release of dexamethasone reduces the amount of inflammation at the site of implantation, which extends the time that the electrode can be used to record brain activity. Other groups obtained similar results when working with polypyrrole electrodes [98].

Another interesting approach to generating electrically controlled release required the fabrication of core-shell fibers where the core consisted of poly(3,4-ethylenedioxythiophene) poly(styrenesulfonate) (PEDOT:PSS) with the shell consisting of polypyrrole [99]. This system was used to deliver bioactive ciprofloxacin, a commonly used antibiotic, when electrical signals were applied. Graphene oxide can also be used for such drug delivery applications [100]. Such electrical controlled delivery systems are now so precise they can deliver cargo to single neurons in vitro [101]. This approach to drug delivery is interesting in the context of electrodes. However, it remains challenging to deliver the appropriate electrical signals necessary to trigger release for the other applications in neural tissue engineering. Dosing also poses an issue when working with such drug delivery systems with potential toxicity to neural tissue being an issue depending on the size of the drug reservoir. These issues amongst other were discussed in a recent review that addresses how to translate such technologies to the clinic [102].

6 OPTICALLY CONTROLLED DRUG DELIVERY SYSTEMS

The use of light provides another interesting avenue for controlling drug release. Such drug delivery systems are photostimulatable and they often use light in the near-infrared wavelength range. These systems have been explored in depth as a way of treating cancers as the use of light stimulation to trigger cargo delivery enables targeted release in the area of the tumor [103]. Such materials could also deliver therapeutics to target regions of the nervous system. In fact, a study used hydrogels loaded with photosensitive polypyrrole nanoparticles that released different factors associated with chemotaxis, including neurontin and semaphorin-3A, to alternatively induce and repel neurite extension from neurons [104], demonstrating the utility of such drug delivery systems. Vitamin B12 can also be used to create tunable drug delivery systems using red, far-red, and near-infrared light [105]. Such systems have been implemented in vitro for neural tissue engineering applications, but remain difficult to implement for in vivo work. However, there is immense potential in terms of using light to control drug release. The recent popularity of optogenetics in neuroscience suggests that figuring out innovative solutions for delivering light cues to the nervous systems will start to be addressed by engineers. The technique of optogenetics uses light to control the function of cells that express light sensitive ion channels and this method has been used to elucidate how different regions of the brain function [106,107]. Optogenetics will be discussed in Chapter 8 as one of the exciting new techniques that has the potential to revolutionize our understanding of the nervous systems [107] and possibly provide new options for therapies for the diseases and disorders of the nervous system [106].

7 FUTURE DIRECTIONS

This chapter has detailed many of the commonly used drug delivery systems for applications in neural tissue engineering along with their associated benefits and drawbacks. It covered the traditional diffusion-based systems and the newer affinity-based delivery systems, which have huge potential in terms of tailoring the release profiles for diverse types of molecules. The use of immobilization to present biomolecules was also discussed in detail along with its disadvantages. The final two types of delivery systems consist of "smart" materials that respond to specific cues (electricity and light) by releasing the target biomolecules. Several promising areas for future work have been mentioned, including more exploration in the use of microfibers as a diffusion-based drug delivery device, the use of encapsulation free affinity-based drug delivery, and further development of stimulus responsive "smart" materials for applications in drug delivery. Other potential cues for such materials include the use of temperature and pH responsive drug delivery systems. Thus, when engineering neural tissue from stem cells, there is a wide variety of potential ways to deliver cues for controlling their behavior. Such systems can also be used to deliver cues to promote regeneration after damage caused by the diseases and disorders of the nervous system in novel fashion and translating these systems for clinical applications will ensure advancement in the area of tissue engineering.

ADDITIONAL RESOURCES

Website for the Controlled Release Society: http://www.controlledreleasesociety.org/Pages/default.aspx

Journal of Controlled Release: http://www.sciencedirect.com/science/journal/01683659

Advanced Drug Delivery Reviews: http://www.sciencedirect.com/science/journal/0169409X

Drug Delivery (Academic Journal): http://www.tandfonline.com/loi/idrd20

BOOKS RELATED TO DRUG DELIVERY

Saltzman, M. Drug Delivery: Engineering Principles for Drug Therapy (Topics in Chemical Engineering). 1st edition. Oxford University Press. 2001.

Ranade, V. and Cannon, J. Drug Delivery Systems. 3rd edition. CRC Press. 2011.

REFERENCES

[1] Elliott Donaghue I, Tam R, Sefton MV, Shoichet MS. Cell and biomolecule delivery for tissue repair and regeneration in the central nervous system. J Control Release 2014;190:219–27.

[2] Russo T, Tunesi M, Giordano C, Gloria A, Ambrosio L. Hydrogels for central nervous system therapeutic strategies. Proc Inst Mech Eng H J Eng Med 2015;229(12):905–16.

[3] Hollis II ER. Axon guidance molecules and neural circuit remodeling after spinal cord injury. Neurotherapeutics 2016;13(2):360–9.

[4] Liechty WB, Kryscio DR, Slaughter BV, Peppas NA. Polymers for drug delivery systems. Annu Rev Chem Biomol Eng 2010;1:149.

[5] Burdick JA, Ward M, Liang E, Young MJ, Langer R. Stimulation of neurite outgrowth by neurotrophins delivered from degradable hydrogels. Biomaterials 2006;27(3):452–9.

[6] Willerth SM, Sakiyama-Elbert SE. Approaches to neural tissue engineering using scaffolds for drug delivery. Adv Drug Deliv Rev 2007;59(4–5):325–38.

[7] Madduri S, Gander B. Schwann cell delivery of neurotrophic factors for peripheral nerve regeneration. J Peripher Nerv Syst 2010;15(2):93–103.

[8] Mann CM, Lee JH, Hillyer J, Stammers AM, Tetzlaff W, Kwon BK. Lack of robust neurologic benefits with simvastatin or atorvastatin treatment after acute thoracic spinal cord contusion injury. Exp Neurol 2010;221(2):285–95.

[9] Sperandeo K, Nogrady L, Moreo K, Prostko CR. Managed approaches to multiple sclerosis in special populations. J Manag Care Pharm 2011;17(9 Suppl C):S1–19. quiz S20–1.

[10] McKeon RJ, Hoke A, Silver J. Injury-induced proteoglycans inhibit the potential for laminin-mediated axon growth on astrocytic scars. Exp Neurol 1995;136(1):32–43.

[11] Morgenstern DA, Asher RA, Fawcett JW. Chondroitin sulphate proteoglycans in the CNS injury response. Prog Brain Res 2002;137:313–32.

[12] Gause I, Sivak WN, Marra KG. The role of chondroitinase as an adjuvant to peripheral nerve repair. Cells Tissues Organs 2015;200(1):59–68.

[13] Hunt D, Coffin R, Anderson P. The Nogo receptor, its ligands and axonal regeneration in the spinal cord; a review. J Neurocytol 2002;31(2):93–120.

[14] Lesny P, De Croos J, Pradny M, Vacik J, Michalek J, Woerly S, et al. Polymer hydrogels usable for nervous tissue repair. J Chem Neuroanat 2002;23(4):243–7.

[15] Drury JL, Mooney DJ. Hydrogels for tissue engineering: scaffold design variables and applications. Biomaterials 2003;24(24):4337–51.

[16] Fisher OZ, Khademhosseini A, Langer R, Peppas NA. Bioinspired materials for controlling stem cell fate. Acc Chem Res 2010;43(3):419–28.

[17] Perale G, Rossi F, Sundstrom E, Bacchiega S, Masi M, Forloni G, et al. Hydrogels in spinal cord injury repair strategies. ACS Chem Neurosci 2011;2(7):336–45.

[18] Asmani MN, Ai J, Amoabediny G, Noroozi A, Azami M, Ebrahimi-Barough S, et al. Three-dimensional culture of differentiated endometrial stromal cells to oligodendrocyte progenitor cells (OPCs) in fibrin hydrogel. Cell Biol Int 2013;37(12):1340–9.

[19] Medberry CJ, Crapo PM, Siu BF, Carruthers CA, Wolf MT, Nagarkar SP, et al. Hydrogels derived from central nervous system extracellular matrix. Biomaterials 2013;34(4):1033–40.

[20] Pakulska MM, Ballios BG, Shoichet MS. Injectable hydrogels for central nervous system therapy. Biomed Mater 2012;7(2):024101.

[21] Lu P, Wang Y, Graham L, McHale K, Gao M, Wu D, et al. Long-distance growth and connectivity of neural stem cells after severe spinal cord injury. Cell 2012;150(6):1264–73.

[22] Lu P, Graham L, Wang Y, Wu D, Tuszynski M. Promotion of survival and differentiation of neural stem cells with fibrin and growth factor cocktails after severe spinal cord injury. J Vis Exp 2014;89:e50641.

[23] Lu P, Woodruff G, Wang Y, Graham L, Hunt M, Wu D, et al. Long-distance axonal growth from human induced pluripotent stem cells after spinal cord injury. Neuron 2014;83(4):789–96.

[24] Freiberg S, Zhu X. Polymer microspheres for controlled drug release. Int J Pharm 2004;282(1–2):1–18.

[25] Zhao C-X. Multiphase flow microfluidics for the production of single or multiple emulsions for drug delivery. Adv Drug Deliv Rev 2013;65(11):1420–46.

[26] Pean JM, Venier-Julienne MC, Boury F, Menei P, Denizot B, Benoit JP. NGF release from poly(D,L-lactide-co-glycolide) microspheres. Effect of some formulation parameters on encapsulated NGF stability. J Control Release 1998;56(1–3):175–87.

[27] Aubert-Pouessel A, Venier-Julienne MC, Clavreul A, Sergent M, Jollivet C, Montero-Menei CN, et al. In vitro study of GDNF release from biodegradable PLGA microspheres. J Control Release Off J Control Release Soc 2004;95(3):463–75.

[28] Agbay A, Mohtaram NK, Willerth SM. Controlled release of glial cell line-derived neurotrophic factor from poly(epsilon-caprolactone) microspheres. Drug Deliv Transl Res 2014;4(2):159–70.

[29] Garbayo E, Montero-Menei CN, Ansorena E, Lanciego JL, Aymerich MS, Blanco-Prieto MJ. Effective GDNF brain delivery using microspheres-a promising strategy for Parkinson's disease. J Control Release 2009;135(2):119–26.

[30] Andrieu-Soler C, Aubert-Pouessel A, Doat M, Picaud S, Halhal M, Simonutti M, et al. Intravitreous injection of PLGA microspheres encapsulating GDNF promotes the survival of photoreceptors in the rd1/rd1 mouse. Mol Vis 2005;11(118–20).

[31] Das S, Carnicer-Lombarte A, Fawcett JW, Bora U. Bio-inspired nano tools for neuroscience. Prog Neurobiol 2016;142:1–22.

[32] Sun B, Taing A, Liu H, Nie G, Wang J, Fang Y, et al. Nerve growth factor-conjugated mesoporous silica nanoparticles promote neuron-like PC12 cell proliferation and neurite growth. J Nanosci Nanotechnol 2016;16(3):2390–3.

[33] Saraiva C, Paiva J, Santos T, Ferreira L, Bernardino L. MicroRNA-124 loaded nanoparticles enhance brain repair in Parkinson's disease. J Control Release 2016;235:291–305.

[34] Singh R, Lillard JW. Nanoparticle-based targeted drug delivery. Exp Mol Pathol 2009;86(3):215–23.

[35] Li J, Mao H, Kawazoe N, Chen G. Insight into the interactions between nanoparticles and cells. Biomater Sci 2016;5(2):173–89.

[36] Madhusudanan P, Reade S, Shankarappa SA. Neuroglia as targets for drug delivery systems: a review. Nanomed Nanotechnol Biol Med 2016;13(2):667–79.

[37] Varela JA, Dupuis JP, Etchepare L, Espana A, Cognet L, Groc L. Targeting neurotransmitter receptors with nanoparticles in vivo allows single-molecule tracking in acute brain slices. Nat Commun 2016;7.

[38] Kumar Bokara K, Suresh Oggu G, Josyula Vidyasagar A, Asthana A, Eun Lee J, Mohan Rao C. Modulation of stem cell differentiation by the influence of nanobiomaterials/carriers. Curr Stem Cell Res Ther 2014;9(6):458–68.

[39] Yi Y, Hsieh I-Y, Huang X, Li J, Zhao W. Glioblastoma stem-like cells: Characteristics, microenvironment, and therapy. Front Pharmacol 2016;7.

[40] Dinda SC, Pattnaik G. Nanobiotechnology-based drug delivery in brain targeting. Curr Pharm Biotechnol 2013;14(15):1264–74.

[41] Maysinger D, Ji J, Hutter E, Cooper E. Nanoparticle-based and bioengineered probes and sensors to detect physiological and pathological biomarkers in neural cells. Front Neurosci 2015;9.

[42] Pandey A, Malek V, Prabhakar V, Aanat Kulkarni Y, Bhanudas GA. Nanoparticles: a neurotoxicological perspective. CNS Neurol Disord Drug Targets 2015;14(10):1317–27 (Formerly Current Drug Targets-CNS & Neurological Disorders).

[43] Kashanian S, Harding F, Irani Y, Klebe S, Marshall K, Loni A, et al. Evaluation of mesoporous silicon/polycaprolactone composites as ophthalmic implants. Acta Biomater 2010;6(9):3566–72.

[44] Lavin DM, Zhang L, Furtado S, Hopkins RA, Mathiowitz E. Effects of protein molecular weight on the intrinsic material properties and release kinetics of wet spun polymeric microfiber delivery systems. Acta Biomater 2013;9(1):4569–78.

[45] Lavin DM, Stefani RM, Zhang L, Furtado S, Hopkins RA, Mathiowitz E. Multifunctional polymeric microfibers with prolonged drug delivery and structural support capabilities. Acta Biomater 2012;8(5):1891–900.

[46] Puhl S, Li L, Meinel L, Germershaus O. Controlled protein delivery from electrospun non-wovens: novel combination of protein crystals and a biodegradable release matrix. Mol Pharm 2014;11(7):2372–80.

[47] Mohtaram NK, Ko J, King C, Sun L, Muller N, Jun MB, et al. Electrospun biomaterial scaffolds with varied topographies for neuronal differentiation of human-induced pluripotent stem cells. J Biomed Mater Res A 2015;103(8):2591–601.

[48] Puppi D, Zhang X, Yang L, Chiellini F, Sun X, Chiellini E. Nano/microfibrous polymeric constructs loaded with bioactive agents and designed for tissue engineering applications: a review. J Biomed Mater Res B Appl Biomater 2014;102(7):1562–79.

[49] Jun Y, Kang E, Chae S, Lee S-H. Microfluidic spinning of micro-and nano-scale fibers for tissue engineering. Lab Chip 2014;14(13):2145–60.

[50] Schaub NJ, Johnson CD, Cooper B, Gilbert RJ. Electrospun fibers for spinal cord injury research and regeneration. J Neurotrauma 2015;33(15):1405–15.

[51] Johnson CD, D'Amato AR, Gilbert RJ. Electrospun fibers for drug delivery after spinal cord injury and the effects of drug incorporation on fiber properties. Cells Tissues Organs 2016;202(1–2):116–35.

[52] Jacobs V, Anandjiwala RD, Maaza M. The influence of electrospinning parameters on the structural morphology and diameter of electrospun nanofibers. J Appl Polym Sci 2010;115(5):3130–6.

[53] Prabaharan M, Jayakumar R, Nair S. Electrospun nanofibrous scaffolds-current status and prospects in drug delivery. In: Biomedical applications of polymeric nanofibers. New York, NY: Springer; 2011. p. 241–62.

[54] Liu W, Thomopoulos S, Xia Y. Electrospun nanofibers for regenerative medicine. Adv Healthcare Mater 2012;1(1):10–25.

[55] Mohtaram NK, Montgomery A, Willerth SM. Biomaterial-based drug delivery systems for the controlled release of neurotrophic factors. Biomed Mater 2013;8(2):022001.

[56] Mansouri N. The influence of topography on tissue engineering perspective. Mater Sci Eng C 2016;61:906–21.
[57] Cheng J, Jun Y, Qin J, Lee S-H. Electrospinning versus microfluidic spinning of functional fibers for biomedical applications. Biomaterials 2016;114:121–43.
[58] Mohtaram N, Ko J, Montgomery A, Carlson M, Sun L, Wong A, et al. Multifunctional electrospun scaffolds for promoting neuronal differentiation of induced pluripotent stem cells. J Biomater Tissue Eng 2014;4:906–14.
[59] Mohtaram N, Ko J, Agbay A, Rattray D, Neill P, Rajwani A, et al. Development of a glial cell-derived neurotrophic factor-releasing artificial dura for neural tissue engineering applications. J Mater Chem B 2015;3(40):7974–85.
[60] Vulic K, Shoichet MS. Affinity-based drug delivery systems for tissue repair and regeneration. Biomacromolecules 2014;15(11):3867–80.
[61] Pakulska MM, Miersch S, Shoichet MS. Designer protein delivery: From natural to engineered affinity-controlled release systems. Science 2016;351(6279):aac4750.
[62] Delplace V, Obermeyer J, Shoichet MS. Local affinity release. ACS Nano 2016;10(7):6433–6.
[63] Taylor SJ, McDonald 3rd JW, Sakiyama-Elbert SE. Controlled release of neurotrophin-3 from fibrin gels for spinal cord injury. J Control Release 2004;98(2):281–94.
[64] Maxwell DJ, Hicks BC, Parsons S, Sakiyama-Elbert SE. Development of rationally designed affinity-based drug delivery systems. Acta Biomater 2005;1(1):101–13.
[65] Willerth SM, Johnson PJ, Maxwell DJ, Parsons SR, Doukas ME, Sakiyama-Elbert SE. Rationally designed peptides for controlled release of nerve growth factor from fibrin matrices. J Biomed Mater Res A 2007;80(1):13–23.
[66] Wood MD, Sakiyama-Elbert SE. Release rate controls biological activity of nerve growth factor released from fibrin matrices containing affinity-based delivery systems. J Biomed Mater Res A 2008;84(2):300–12.
[67] Fu A, Thatiparti T, Saidel G, Recum H. Experimental studies and modeling of drug release from a tunable affinity-based drug delivery platform. Ann Biomed Eng 2011;39(9):2466–75.
[68] Oduah EI, Linhardt RJ, Sharfstein ST. Heparin: Past, present, and future. Pharmaceuticals 2016;9(3):38.
[69] Sakiyama-Elbert SE. Incorporation of heparin into biomaterials. Acta Biomater 2014;10(4):1581–7.
[70] Sakiyama SE, Schense JC, Hubbell JA. Incorporation of heparin-binding peptides into fibrin gels enhances neurite extension: an example of designer matrices in tissue engineering. FASEB J 1999;13(15):2214–24.
[71] Sakiyama-Elbert SE, Hubbell JA. Controlled release of nerve growth factor from a heparin-containing fibrin-based cell ingrowth matrix. J Control Release 2000;69(1):149–58.
[72] Sakiyama-Elbert SE, Hubbell JA. Development of fibrin derivatives for controlled release of heparin-binding growth factors. J Control Release 2000;65(3):389–402.
[73] Lee AC, Yu VM, Lowe 3rd JB, Brenner MJ, Hunter DA, Mackinnon SE, et al. Controlled release of nerve growth factor enhances sciatic nerve regeneration. Exp Neurol 2003;184(1):295–303.
[74] Taylor SJ, Sakiyama-Elbert SE. Effect of controlled delivery of neurotrophin-3 from fibrin on spinal cord injury in a long term model. J Control Release Off J Control Release Soc 2006;116(2):204–10.

[75] Taylor SJ, Rosenzweig ES, McDonald 3rd JW, Sakiyama-Elbert SE. Delivery of neurotrophin-3 from fibrin enhances neuronal fiber sprouting after spinal cord injury. J Control Release 2006;113(3):226–35.
[76] Wood MD, Moore AM, Hunter DA, Tuffaha S, Borschel GH, Mackinnon SE, et al. Affinity-based release of glial-derived neurotrophic factor from fibrin matrices enhances sciatic nerve regeneration. Acta Biomater 2009;5(4):959–68.
[77] Wood MD, Borschel GH, Sakiyama-Elbert SE. Controlled release of glial-derived neurotrophic factor from fibrin matrices containing an affinity-based delivery system. J Biomed Mater Res A 2009;89A(4):909–18.
[78] Willerth SM, Rader A, Sakiyama-Elbert SE. The effect of controlled growth factor delivery on embryonic stem cell differentiation inside fibrin scaffolds. Stem Cell Res 2008;1(3):205–18.
[79] Johnson PJ, Tatara A, Shiu A, Sakiyama-Elbert SE. Controlled release of neurotrophin-3 and platelet-derived growth factor from fibrin scaffolds containing neural progenitor cells enhances survival and differentiation into neurons in a subacute model of SCI. Cell Transplant 2010;19(1):89–101.
[80] Wilems TS, Pardieck J, Iyer N, Sakiyama-Elbert SE. Combination therapy of stem cell derived neural progenitors and drug delivery of anti-inhibitory molecules for spinal cord injury. Acta Biomater 2015;28:23–32.
[81] Martino MM, Briquez PS, Ranga A, Lutolf MP, Hubbell JA. Heparin-binding domain of fibrin (ogen) binds growth factors and promotes tissue repair when incorporated within a synthetic matrix. Proc Natl Acad Sci 2013;110(12):4563–8.
[82] Ho Y-C, Mi F-L, Sung H-W, Kuo P-L. Heparin-functionalized chitosan–alginate scaffolds for controlled release of growth factor. Int J Pharm 2009;376(1):69–75.
[83] Ma F, Xiao Z, Chen B, Hou X, Dai J, Xu R. Linear ordered collagen scaffolds loaded with collagen-binding basic fibroblast growth factor facilitate recovery of sciatic nerve injury in rats. Tissue Eng A 2014;20(7–8):1253–62.
[84] Cao J, Sun C, Zhao H, Xiao Z, Chen B, Gao J, et al. The use of laminin modified linear ordered collagen scaffolds loaded with laminin-binding ciliary neurotrophic factor for sciatic nerve regeneration in rats. Biomaterials 2011;32(16):3939–48.
[85] Fan J, Xiao Z, Zhang H, Chen B, Tang G, Hou X, et al. Linear ordered collagen scaffolds loaded with collagen-binding neurotrophin-3 promote axonal regeneration and partial functional recovery after complete spinal cord transection. J Neurotrauma 2010;27(9):1671–83.
[86] Han Q, Sun W, Lin H, Zhao W, Gao Y, Zhao Y, et al. Linear ordered collagen scaffolds loaded with collagen-binding brain-derived neurotrophic factor improve the recovery of spinal cord injury in rats. Tissue Eng A 2009;15(10):2927–35.
[87] Vulic K, Shoichet MS. Tunable growth factor delivery from injectable hydrogels for tissue engineering. J Am Chem Soc 2011;134(2):882–5.
[88] Pakulska MM, Vulic K, Shoichet MS. Affinity-based release of chondroitinase ABC from a modified methylcellulose hydrogel. J Control Release 2013;171(1):11–6.
[89] Parker J, Mitrousis N, Shoichet MS. Hydrogel for simultaneous tunable growth factor delivery and enhanced viability of encapsulated cells in vitro. Biomacromolecules 2016;17(2):476–84.
[90] Dutta D, Fauer C, Mulleneux H, Stabenfeldt S. Tunable controlled release of bioactive SDF-1α via specific protein interactions within fibrin/nanoparticle composites. J Mater Chem B 2015;3(40):7963–73.
[91] Pakulska MM, Donaghue IE, Obermeyer JM, Tuladhar A, McLaughlin CK, Shendruk TN, et al. Encapsulation-free controlled release: electrostatic adsorption eliminates the need for protein encapsulation in PLGA nanoparticles. Sci Adv 2016;2(5):e1600519.

[92] Meade KA, White KJ, Pickford CE, Holley RJ, Marson A, Tillotson D, et al. Immobilization of heparan sulfate on electrospun meshes to support embryonic stem cell culture and differentiation. J Biol Chem 2013;288(8):5530–8.

[93] Lee M, Kim Y, Ryu JH, Kim K, Han Y-M, Lee H. Long-term, feeder-free maintenance of human embryonic stem cells by mussel-inspired adhesive heparin and collagen type I. Acta Biomater 2016;32:138–48.

[94] McCormick AM, Jarmusik NA, Leipzig ND. Co-immobilization of semaphorin3A and nerve growth factor to guide and pattern axons. Acta Biomater 2015;28:33–44.

[95] Stefonek-Puccinelli TJ, Masters KS. Co-immobilization of gradient-patterned growth factors for directed cell migration. Ann Biomed Eng 2008;36(12):2121–33.

[96] Wang TY, Bruggeman KA, Sheean RK, Turner BJ, Nisbet DR, Parish CL. Characterization of the stability and bio-functionality of tethered proteins on bioengineered scaffolds: implications for stem cell biology and tissue repair. J Biol Chem 2014;289(21):15044–51.

[97] Wadhwa R, Lagenaur CF, Cui XT. Electrochemically controlled release of dexamethasone from conducting polymer polypyrrole coated electrode. J Control Release 2006;110(3):531–41.

[98] Leprince L, Dogimont A, Magnin D, Demoustier-Champagne S. Dexamethasone electrically controlled release from polypyrrole-coated nanostructured electrodes. J Mater Sci Mater Med 2010;21(3):925–30.

[99] Esrafilzadeh D, Razal JM, Moulton SE, Stewart EM, Wallace GG. Multifunctional conducting fibres with electrically controlled release of ciprofloxacin. J Control Release 2013;169(3):313–20.

[100] Weaver CL, LaRosa JM, Luo X, Cui XT. Electrically controlled drug delivery from graphene oxide nanocomposite films. ACS Nano 2014;8(2):1834–43.

[101] Kim T-H, Cho H-Y, Lee K-B, Kim SU, Choi J-W. Electrically controlled delivery of cargo into single human neural stem cell. ACS Appl Mater Interfaces 2014;6(23):20709–16.

[102] Asplund M, Boehler C, Stieglitz T. Anti-inflammatory polymer electrodes for glial scar treatment: bringing the conceptual idea to future results. The chronic challenge-new vistas on long-termmultisite contacts to the central nervous system. Front Neuroeng 2014;7(9):46.

[103] Kim H, Chung K, Lee S, Kim DH, Lee H. Near-infrared light-responsive nanomaterials for cancer theranostics. Wiley Interdiscipl Rev Nanomed Nanobiotechnol 2016;8(1):23–45.

[104] Li W, Luo R, Lin X, Jadhav AD, Zhang Z, Yan L, et al. Remote modulation of neural activities via near-infrared triggered release of biomolecules. Biomaterials 2015;65:76–85.

[105] Shell TA, Lawrence DS. Vitamin B12: A tunable, long wavelength, light-responsive platform for launching therapeutic agents. Acc Chem Res 2015;48(11):2866–74.

[106] Song C, Knöpfel T. Optogenetics enlightens neuroscience drug discovery. Nat Rev Drug Discov 2015;15(2):97–109.

[107] Deisseroth K. Optogenetics: 10 years of microbial opsins in neuroscience. Nat Neurosci 2015;18(9):1213–25.

CHAPTER

New technologies for engineering neural tissue from stem cells

8

1 INTRODUCTION TO NEW TECHNOLOGIES FOR ENGINEERING NEURAL TISSUES FROM STEM CELLS

This book has covered a significant amount of information, laying a solid foundation for applying the traditional tissue engineering principles as a way of generating functional substitutes for diseased or damaged neural tissue. Chapters 1–7 covered these topics in an in-depth level of detail. This chapter discusses several new and emerging strategies for engineering neural tissue that represent a dynamic shift from this existing paradigm. Several exciting new technologies have been developed that could provide major breakthroughs in the field of neural tissue engineering as well as for neuroscience in general. In particular, this chapter will explore recent developments in biotechnology, including the generation of neural organoids, the use of 3D bioprinting for neural tissue engineering applications, direct reprogramming of cells into neural tissue, gene editing to correct neurological disorders, and the technology known as optogenetics where light is used to control the function of neural cells, as well as the use of its chemical counterpart—Designer Receptors Exclusively Activated by Designer Drugs (DREADDs). Overall, this set of technologies provides insight into how the field of neural tissue engineering will evolve in the future.

2 GENERATION OF NEURAL ORGANOIDS FROM STEM CELLS

This book has covered the properties and potential of stem cells, along with how they can be combined with biomaterial scaffolds to engineer neural tissue. Chapter 3 discussed how one common method for promoting the differentiation of pluripotent stem cells requires the formation of aggregates known as embryoid bodies [1]. This aggregation process mimics the events that occur during embryogenesis, causing the pluripotent stem cells to mature into differentiated phenotypes. Often times the cells that arise from embryoid bodies are sorted out by progenitor type before being subjected to more extensive differentiation protocols. Recently, focus has been shifted to using the inherent properties of pluripotent stem cells that enable them to self-assemble into functional tissues under appropriate conditions [2]. These functional mini-tissues are known as organoids. These organoids serve as mini-models of tissues for research applications as well as important tools for understanding the process of development. They can be

Engineering Neural Tissue from Stem Cells. http://dx.doi.org/10.1016/B978-0-12-811385-1.00008-X

fabricated from human-induced pluripotent stem cell lines, which enable replication of disease phenotypes present in the patient where the initial cells were sourced [3–9].

One of the recent landmark studies in engineering neural tissue organoids occurred when Dr. Madeline Lancaster (now at the University of Cambridge) and Dr. Juergen Knoblich at Institute of Molecular Biotechnology in Vienna showed that human pluripotent stem cells could be used to engineer "mini-brains" in a dish by harnessing the inherent properties of these cells (Fig. 1) [5]. While previous work had shown that mouse embryonic stem cells could generate organoids that resembled the optic cup found in the eye [10], this study was the first to use human pluripotent stem cells to replicate the anatomical structures found in brain. However, these tissues did not grow to their normal size even after 10 months of culture. It is possible that the size of organoids was restricted due to the lack of blood vessels. These organoids serve as valuable tools in neurobiology as they can be used to understand how development of the brain occurs, along with how different diseases affect this process. They also serve as effective tools for screening drugs to determine their potential toxicity and efficacy [11].

For example, the outbreak of the Zika virus that began in 2015 created an urgent need to understand how it affected the nervous system in order to develop an effective treatment. This virus induces microcephaly, along with other brain defects, in infants. In adults, infection with Zika virus leads to Guillain-Barré syndrome, where the peripheral nervous system becomes damaged, leading to muscle weakness [12]. Recent work on brain-derived organoids has focused on using them as a tool to understand the Zika virus, as infecting them with the virus replicated the effect on brain size postinfection, demonstrating that the virus inhibits proliferation of neural progenitors [8,13–15]. This body of work illustrates how important these organoids are as a tool for studying how human tissues respond to disease, as well as how they can provide insight into the mechanisms of such diseases. As mentioned in earlier in this section, there is increased interest in making these organoids more closely mimic the tissues and their properties found in the nervous system. Being able to include the vasculature in brain organoids would represent a major step forward for this technology as it would enable the generation of larger organoids that match the human brain more closely in terms of size. More work will also be necessary to determine the range of diseases that can be replicated using this novel brain organoid technology.

A recent study used pluripotent stem cells to generate organoids that resemble the cornea [16]. Another goal for organoid technology includes being able to fully replicate the structures found in the eye, including the retina, using human pluripotent stem cells. These organoids would be useful for modeling the different diseases and disorders that afflict vision, and they could contribute toward finding potential cures [17]. Finally, while there has been some success in generating organoids that resemble the brain and eye, no major attempts have been made for replicating the spinal cord in a dish. Such organoids would also be of high interest to the research community. Overall, organoids serve as a highly promising technology for applications in neural tissue engineering, and as an important tool for identifying novel therapies for the diseases and disorders of the central nervous system.

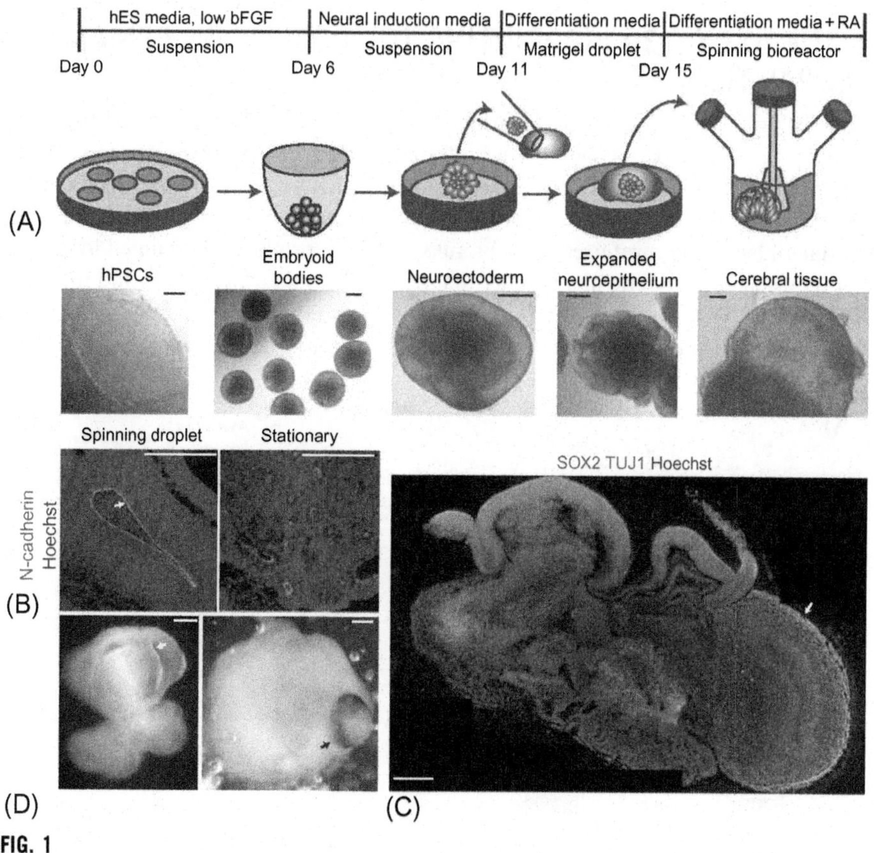

FIG. 1

Description of cerebral organoid culture system. (A) Schematic of the culture system described in detail in methods. Example images of each stage are shown.
(B) Neuroepithelial tissues generated using this approach (left panel) exhibited large fluid-filled cavities and typical apical localization of the neural N-cadherin *(arrow)*. These tissues were larger and more continuous than tissues grown in stationary suspension without Matrigel (right panel). (C) Sectioning and immunohistochemistry revealed complex morphology with heterogeneous regions containing neural progenitors (Sox2, *red in the electronic version, gray in print version*) and neurons (Tuj1, *green in the electronic version, gray in print version*) *(arrow)*. (D) Low magnification bright field images revealing fluid-filled cavities reminiscent of ventricles *(white arrow)* and retina tissue, as indicated by retinal pigmented epithelium *(black arrow)*. Scale bars: 200 μm.

Reprinted with permission from Lancaster MA, Renner M, Martin C-A, Wenzel D, Bicknell LS, Hurles ME, et al. Cerebral organoids model human brain development and microcephaly. Nature 2013;501(7467):373–9.

3 THE USE OF 3D BIOPRINTING TO PRODUCE ON DEMAND NEURAL TISSUES

3D printing is a process where materials are added layer by layer to construct a 3D object [18]. This process can also be referred to as additive manufacturing. In the case of 3D bioprinting, these materials include biomaterials and cells, which can be printed into constructs that resemble the tissues found in the body. These printable biomaterials are often referred to as bioinks and their properties are tuned to ensure that they can deliver cells in a manner that retains both their properties and function [19]. A variety of different 3D bioprinters exist for fabricating engineered tissues. Fig. 2 shows a selection of commonly used bioprinting techniques, including (1)

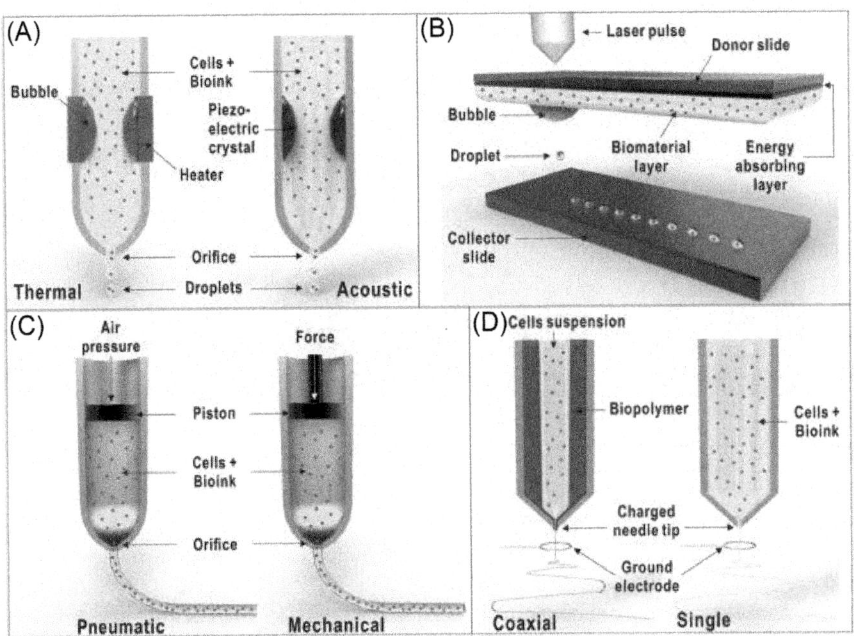

FIG. 2

Different types of bioprinting techniques used for fabricating cell-laden constructs. (A) Inkjet bioprinting method: bubbles created through thermal heaters or bioink forced out through piezoelectric crystal. (B) Laser-assisted bioprinting method: light energy from the laser source creates bubbles in biomaterial layer in the ribbon (donor slide/energy absorbing layer) and droplets of biomaterial are deposited on the collector slide. (C) Extrusion bioprinting method: uses syringe with piston that moves via air pressure or mechanical force to expel bioink through the orifice in continuous stream or droplets depending on the viscosity of material. (D) Bio-electrospraying/cell electrospinning: it uses charged need tip at orifice to accelerate bioink and expel them in a controllable manner.

Reprint with permission from Hong N, Yang GH, Lee J, Kim G. 3D bioprinting and its in vivo applications. J Biomed Mater Res B Appl Biomater 2017.

inkjet-based printing, (2) laser-assisted printing, (3) extrusion-based printing, and (4) cell electrospinning. These techniques differ in how they deliver the bioinks to create 3D structures. Such 3D bioprinting offers two major advantages over traditional tissue culture. It can reduce the amount of labor needed to generate such tissues by automating part of the tissue production process, which in turn decreases the possibility of contamination. 3D bioprinting can also generate such tissues in a high throughput manner in comparison to traditional tissue engineering techniques.

Inkjet printing methods deliver droplets of ink onto a surface through a combination of pressure-driven flow and electrostatic forces. Often commercially available inkjet printers are modified to enable the delivery of bioinks, as well as to add a–z dimension, enabling the fabrication of 3D objects. In 2006, the Boland group at Clemson University used inkjet printing to layer primary neural cells, consisting of hippocampal and cortical embryonic neurons, into functional 3D neural tissues as assessed by electrophysiology [20]. In 2009, another group used an extrusion-based bioprinter to fabricate neural tissues by printing rat embryonic neural stem cells, including neurons and astrocytes, contained within a collagen bioink whose polymerization was induced by addition of sodium biocarbonate [21]. They were able to print these cells into rings and cross-like structures consisting of multiple layers and these cells expressed neuronal and astrocytic markers after 15 days of culture. This report was the first to demonstrate the generation of such novel multilayer composite neural tissues. A study performed the following year showed that a novel combination of collagen bioink along with fibrin bioink to generate controlled release of vascular endothelial growth factor (VEGF) resulted in the construction of 3D neural tissues from mouse neural progenitors [22]. Recent work from the Feinberg group at Carnegie Mellon developed a novel 3D printing technique based on freeform reversible embedding of suspended hydrogels (FRESH), including a brain-like structure [23]. This advancement in bioprinting allows for bioinks made from soft proteins and polysaccharides as these materials are printed under special conditions to ensure their stability.

3D bioprinting possesses several properties that make it attractive for constructing conduits for repairing damage to peripheral nerves. These techniques can fabricate conduits that contain channels for guiding cell growth, and varying the orientation, size, and geometry of these channels influences how cells behave when migrating through such engineered tissues [24]. For example, laser-based microstereolithography was used to successfully produce 3D nerve guidance conduits with defined topography by using a photocurable version of poly(ethylene glycol) [25]. These conduits supported nerve regeneration over a 3 mm gap in vivo after 21 days. More recent work has also used 3D bioprinting methods to produce different varieties of nerve guidance conduits [26–28]. Another group used a novel bioink consisting of the polysaccharide gellan gum functionalized with the RGD peptide to print cortical neurons into tissue resembling that found in the brain [29]. Other groups have investigated the use of bioinks composed of polyurethane as well as ones composed of different polysaccharide blends for printing neural progenitors [30,31].

Areas for future work include applying the variety of techniques discussed in this section for bioprinting neural tissues derived from pluripotent stem cells. In addition, the development of optimized bioinks will enable printing of more functionally relevant engineered tissues that replicate the features of the central and peripheral nervous system. Intriguingly, such systems can print thermoresponsive hydrogels for cell-culture applications, which could enable the production of stimulus responsive neural tissues that change their behavior in response to temperature [32]. 3D printing also enables fabrication of engineered tissues with specific dimensions and properties based on the contents of the starting file, which could lead to printing personalized neural tissues on demand as done for other tissues types by the Vancouver-based start-up Aspect Biosystems [33]. These tissues also can be used as a tool for screening potential drugs in addition to being used as replacement tissues.

4 IN VITRO AND IN VIVO ENGINEERING OF NEURAL TISSUE FROM SOMATIC CELLS

The development of induced pluripotent stem cells provided a powerful example of how cells can be transformed into an entirely different phenotype by genetic manipulation of transcription factor expression levels [34–36] and as was touched on briefly in Chapter 3. This reprogramming process used viral methods to express transcription factors associated with embryonic stem cells in mature cells, causing them to revert to a pluripotent state through a process referred to as reprogramming. Such viral-mediated reprogramming methods can also directly convert differentiated cells into other mature cell types, including neurons as seen in Fig. 3 [37–47]. A landmark 2010 study from Dr. Marius Wernig's group performed a similar screening process to the one used to generate induced pluripotent stem cells to determine which

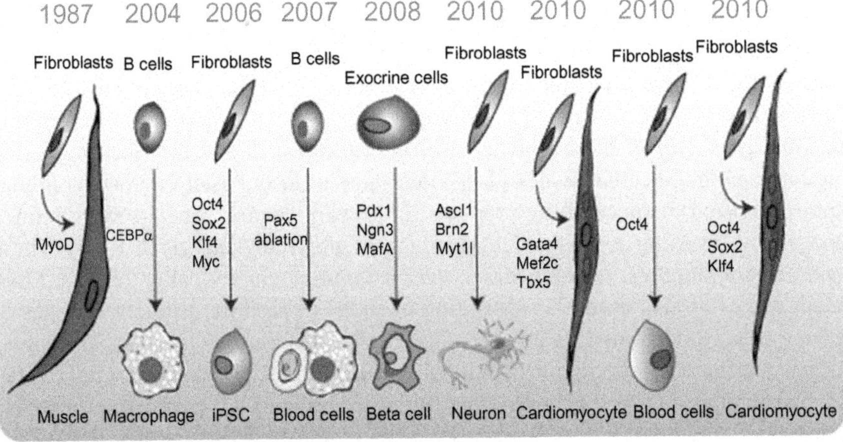

FIG. 3

The timeline of different discoveries of how different transcription factors can convert one cell phenotype to another.

Reprinted with permission from Systems Bioscience Inc.

transcription factors could directly reprogram fibroblasts cultured in vitro into neuronal like cells referred to as induced neurons [37]. This study identified a combination of three transcription factors associated with the developing nervous system—Brn2, Ascl1, and Mytl1—whose overexpression resulted in the reprogramming of mouse fibroblasts into induced neurons. This combination of transcription factors is referred to as the BAM factors. These cells, shown in Fig. 4, express neuronal markers and can fire action potentials, indicating their functionality.

Of particular interest has been the transcription factor Ascl1 as it serves as a powerful regulator of developmental neurogenesis, where it controls the proliferation and differentiation of neural stem cells [48]. Ascl1 binds condensed chromatin, enabling it to recruit other transcription factors that determine neural cell fate [49]. This potent reprogramming ability, along with its ability to recruit other transcription factors, established Ascl1 as a "pioneer" transcription factor. In 2014, the Wernig group directly converted fibroblasts into neurons by transfecting these cells using only a lentiviral vector encoding Ascl1 [45]. The conversion efficiency of the reprogramming process was ~10%.

More recent work has demonstrated that the most effective methods for reprogramming human fibroblasts into neurons require a combination of transcription factors, such as the BAM factors [37] and Lmx1a, Ascl1, and Nurr1 (known as LAN factors) [50]. Human fibroblasts also possess a lower reprogramming efficiency compared to mouse fibroblasts when using virus-mediated transcription factor expression [51]. Other in vitro engineering approaches to generating neural tissue from somatic cells include the expression of microRNAs using viral methods to reprogram human fibroblasts into functional neurons [52–54]. Recent work from Duke University showed that CRISPR/Cas9 technology could target expression of the BAM factors, resulting in the reprogramming of mouse fibroblasts into neurons [55]. Section 5 discusses the use of such gene-editing technologies in an in-depth manner. As an alternative to viral-mediated methods, several groups have used small molecule cocktails to convert fibroblasts into different types of cells found in the nervous system, including neural stem cells, astrocytes (neural support cells), and neurons [56–60].

The concept of reprogramming one mature cell type into another has also been explored as a way of engineering desired tissues in vivo. This process of in vivo reprogramming avoids several of the major issues associated with cell transplantation, while providing a pathway for the next generation of regenerative medicine strategies [61]. Direct conversion of a patient's own cells into the cells necessary for restoring lost function eliminates the need for immunosuppression and invasive surgeries associated with traditional cell transplantation therapies. Therapies relying on direct reprogramming could translate more readily for clinical applications as they avoid the regulatory testing associated with transplanting cells, especially the need to ensure the safety of transplanted pluripotent stem cell-derived products by confirming that no tumors form postimplantation. Such methods also do not require the in vitro culture of sufficient quantities of cells for transplantation, which can be a costly and labor-intensive process. In addition, it is often a rate limiting step for how many patients can receive certain cell therapies.

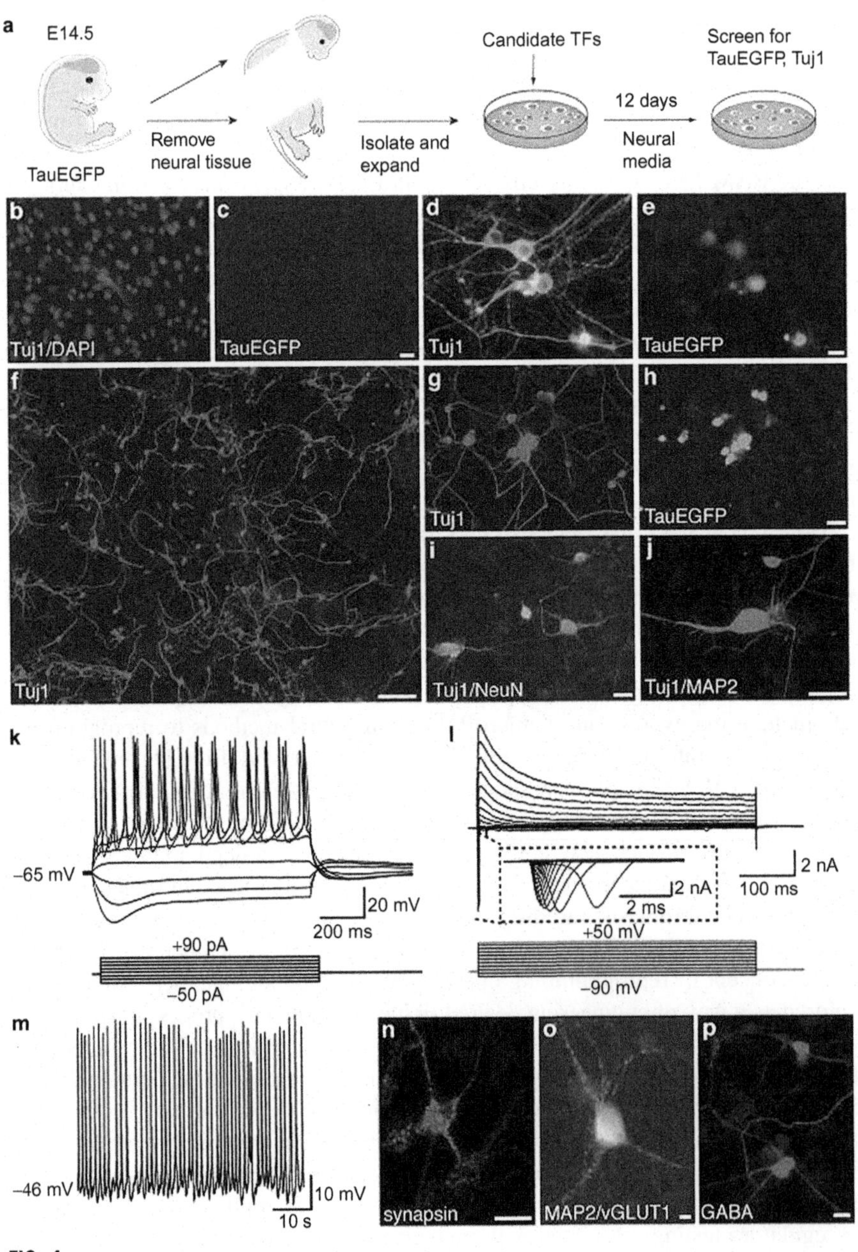

FIG. 4

See the legend on opposite page

In vivo reprogramming using viral-mediated transcription factor expression can successfully engineer functional cardiac tissue in preclinical disease models, suggesting the viability of in vivo reprogramming for tissue engineering applications [62,63]. In more relevant work, several groups have converted various cell types in vivo into desirable phenotypes for engineering neural tissue [47,64–68]. Niu *et al.* showed that (1) the viral expression of the transcription factor Sox2 converted astrocytes in the brain into neural progenitors that matured into functional neurons [47], and (2) the overexpression of the transcription factor NeuroD converted reactive astrocytes present after brain injury into neurons [64]. This work demonstrates how the transcription factors associated different neural development stages can initiate reprogramming into the different cellular phenotypes found in the nervous system. Another study demonstrated that both human fibroblasts and astrocytes can be reprogrammed into neurons using similar methods [65]. Follow-up work confirmed that these reprogrammed cells integrated into the neural circuitry of the brain, suggesting the generation of functional tissue [66]. In 2015, Liu and colleagues used adeno-associated virus (AAV) to express Ascl1 in astrocytes found in the brain, turning these cells into functional neurons [67]. These previous examples show that in vivo neural tissue engineering can occur in the brain. Successful cellular reprogramming can be achieved in the injured spinal cord as well, suggesting that this technique can be applied to the other major organ present in the central nervous system for in vivo engineering applications. For example, viral overexpression of Sox2 in combination with valproic acid treatment converted the astrocytes present in the injured spinal cord into neural progenitors [68]. These neural

FIG. 4

A screen for neuronal fate inducing factors and characterization of MEF-derived induced neuron (iN) cells. (A) Experimental rationale. (B) Uninfected, p3 TauEGFP mouse embryonic fibroblasts (MEFs) contained rare Tuj1-positive cells *(red)* with flat morphology. *Blue*: DAPI counterstain. (C) Tuj1-positive fibroblasts do not express visible TauEGFP. (D and E) MEF-iN cells express Tuj1 (neuronal marker, *red*) and TauEGFP *(green)* and display complex neuronal morphologies 32 days after infection with the 19-factor (19F) pool. (F) Tuj1 expression in MEFs 13 days after infection with the 5F pool. (G–J) MEF-derived Tuj1-positive iN cells co-express the pan-neuronal markers TauEGFP (H), NeuN (*red in the electronic version, gray in print version,* I) and MAP2 (*red in the electronic version, gray in print version,* J). (K) Representative traces of membrane potential responding to step depolarization by current injection (lower panel). Membrane potential was current-clamped at around −65 mV. (L) Representative traces of whole-cell currents in voltage-clamp mode, cell was held at −70 mV, step depolarization from −90 to 60 mV at 10 mV interval were delivered (lower panel). Insert showing Na+ currents. (M) Spontaneous action potentials (AP) recorded from a 5F MEF-iN cell 8 days postinfection. No current injection was applied. (N–P) 22 days postinfection 5F MEF-iN cells expressed synapsin (*red,* N) and vesicular glutamate transporter 1 (vGLUT1) (*red,* O) or GABA (P). Scale bars = 5 μm (O), 10 μm (E,N,P) 20 μm (C,H,I), and 200 μm (F).

Reprinted with permission from Vierbuchen T, Ostermeier A, Pang ZP, Kokubu Y, Sudhof TC, Wernig M. Direct conversion of fibroblasts to functional neurons by defined factors. Nature 2010;463(7284):1035–41.

progenitors then differentiated into mature neurons, providing a potential mechanism for promoting functional recovery post-injury.

While these examples reflect the promise of reprogramming mature cells into neural phenotypes for applications both in vitro and in vivo, there are still limitations on these processes. Using viral vectors to generate protein expression remains problematic for clinical applications. The process alters the genome of the infected cells while producing inconsistent levels of protein expression due to heterogeneous infection rates of the cells present [69–71]. The use of small molecule cocktails avoids some of these issues, but these cocktails are often supplemented with growth factors to achieve successful reprogramming. Other major limitations associated with small molecule cocktails include achieving proper dosing and the potential for off-target effects in vivo [72]. My research group recently validated a novel fusion version of the transcription factor Ascl1 called IASCL1 as powerful controller of stem cell differentiation into neurons and we are currently exploring if IASCL1 can reprogram fibroblasts and astrocytes into neurons [73]. The use of functionalized transcription factors provides an attractive alternative to using viral vectors and small molecules for reprogramming as detailed in a commentary from our lab [74]. It avoids the inconsistent expression levels associated with viral methods and reduces the potential for off target effects associated with small molecule cocktails. Overall, direct reprogramming serves as a powerful tool for engineering neural tissue.

5 GENOME EDITING FOR ENGINEERING NEURAL TISSUE

The technique of genome/gene editing consists of the process of changing the genome of a cell, which was selected by Nature Biotechnology as the Method of the Year in 2011 [75]. These changes to the genome include additions, deletions, or editing of DNA segments to correct mutations, which can serve as a powerful tool for applications in neural tissue engineering. The purpose of editing a genome is to have a significant impact on the resulting functions of the cell that was manipulated. For example, being able to accurately correct the genetic mutation associated with Huntington's disease could lead to a possible cure by restoring proper neuronal survival and function [76].

This section focuses on tools for editing the genome that rely on enzymes known as nucleases that cleave DNA to create double stranded breaks, which are regions where both strands of DNA are broken—leaving a gap. The area where the break occurs is then repaired with the resulting DNA sequence being manipulated based on the desired DNA sequence. Fig. 5 shows three of the most commonly used nuclease-based systems for genome editing. Initially, such editing was accomplished by inducing a double-stranded break along with DNA corresponding to the region and relying the endogenous repair mechanism of homologous recombination to introduce the target sequence in the genome. Variations on this process can also result in targeted deletions of a DNA region, as well as introduction or correction of mutations in the DNA sequence. While the initial attempts at gene editing were highly nonspecific, significant improvements have been made to this process in recent years.

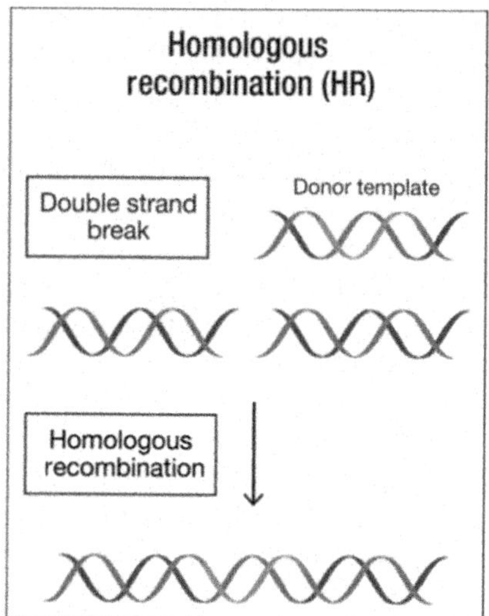

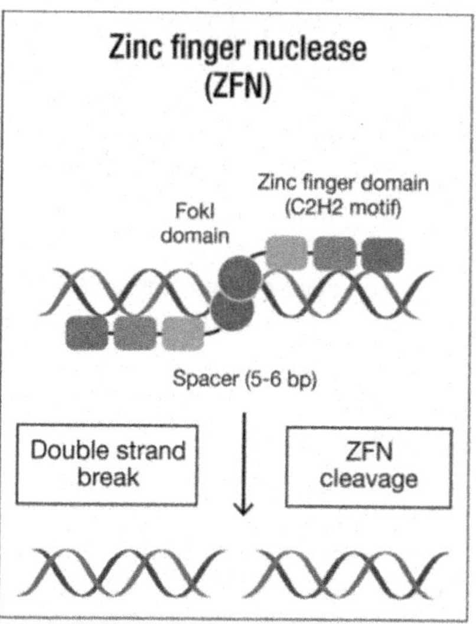

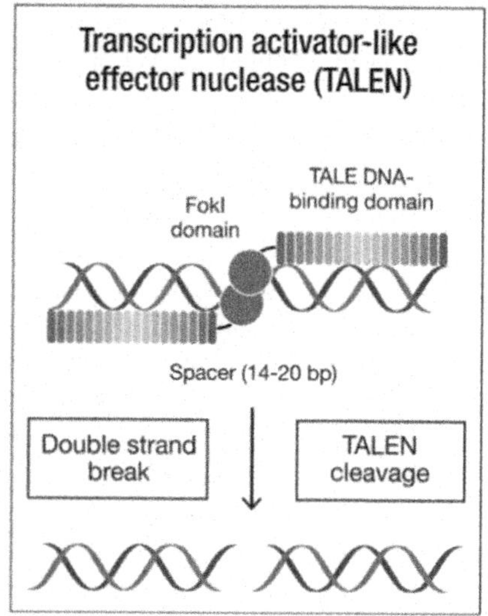

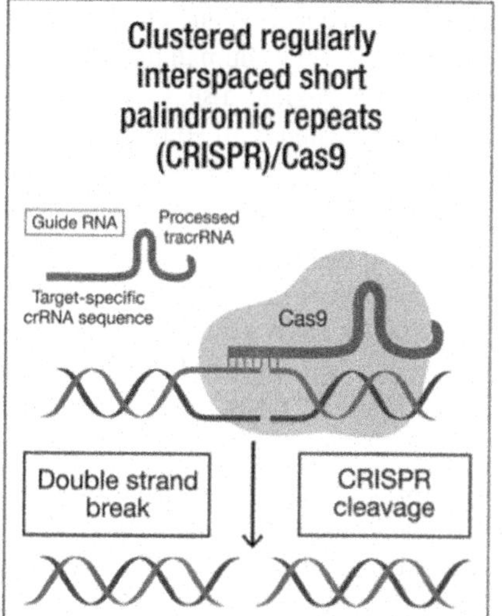

FIG. 5

Genome-editing methodologies which can be applied to human pluripotent stem cells. Homologous recombination (HR), or the more advanced tools such as zinc finger nucleases (ZFNs), transcription activator-like effector nucleases (TALENs) or clustered regularly interspaced short palindromic repeat (CRISPR)/Cas system can be applied to human pluripotent stem cells (hPSCs) either to (1) create naturally occurring mutations or (2) repair a mutation to generate isogenic controls in hPSCs, to understand the function of a gene of interest.

Reprinted with permission from Teo AKK, Gupta MK, Doria A, Kulkarni RN. Dissecting diabetes/metabolic disease mechanisms using pluripotent stem cells and genome editing tools. Mol Metabolism 2015;4(9):593–604.

A major breakthrough in gene editing came when scientists decided to engineer novel artificial nucleases consisting of the cleavage domain of *Fok* I endonuclease fused to DNA targeting domains called zinc finger motifs, along with the desired cleavage sequence [77]. These zinc finger domains enable targeting of specific genome regions where a double-strand break is induced and the original DNA sequence can be manipulated as described above. Zinc finger domains represent one of the most common DNA-binding motifs present in the human genome, making it highly versatile for gene-editing applications. It is possible to order targeted zinc finger endonucleases for specific DNA sequences and their associated proteins from various companies such as Sangamo Therapeutics and Sigma-Aldrich, whose websites are given in the resource list at the end of the chapter. While the development of zinc finger nucleases represented a major step in targeted genome editing, these enzymes still had off target effects as they can cleave regions of DNA that were not the desired target sequence. The use of the Fok1 endonuclease was also problematic as it is derived from a bacterial cell, and thus it could induce a possible immune response upon injection in the human body. Both unintended side-effects made it difficult to translate therapies based on this technology for clinically relevant genome editing. Despite these concerns, one clinical trial has successfully used zinc finger nucleases to treat a rare form of human immunodeficiency virus in a limited number of patients [78]. For neurological diseases, zinc finger nucleases have been used to correct the mutation associated with Huntington's disease in preclinical work involving a mouse disease model [79].

While this technology represented a major advance in genome editing, it still suffered from a lack of specificity when targeting regions of DNA for editing. Thus, the next generation of gene-editing tools were created with a higher degree of specificity for these target DNA sequences. These tools were termed transcription activator-like effector-based nucleases (TALENs), which were easier to develop in terms of the potential range of DNA sequences they could edit [80]. In this case, domains taken from proteins expressed by *Xanthomonas* bacteria designed to infiltrate the immune systems of plants were selected based on their ability to bind target sequences of DNA. These sequences were adapted for use in mammalian cells where they could be used in a manner similar to how zinc finger proteins bind specific regions of DNA, marking them for cleavage [81]. These bacterial protein sequences provided a more accurate method of editing genomes, although there were still issues about potential immunogenicity of these constructs as well as worries about potential off target effects. TALEN technology also opened up the number of genes that could be targeted as they have a specific DNA-protein code where specific domains target each unique nucleotide present in DNA, making it easier to design the desired protein-binding sequence when compared to using zinc finger protein domains to target DNA.

Finally, one of the most recent gene-editing technologies relies on the use of clustered regularly interspaced short palindromic repeats (CRISPR) and the Cas9 protein to target and alter the genome of mammalian cells [82–84]. It was named the 2015 Breakthrough of the Year by the journal Science and Jennifer Lopez is producing a television show called CRISPR, indicating the high level of public interest in this technology. In this gene-editing system, guide molecules consisting of RNA are used to target the desired regions of DNA for editing as opposed to using a designer protein as is done with zinc finger technology and TALENS. The RNA sequence is selected based on the target DNA region instead of having to manipulate an amino acid sequence to achieve specificity. Thus, this system is easy to adapt for a wider range of target sequences compared to the previous gene-editing strategies that used protein domains. While Dr. Jennifer Doudna (University of California-Berkeley) along with colleagues including Dr. Emmanuelle Charpentier (Max Planck Institute for Infection Biology) were the first to adapt the CRIPSR/Cas9 as a tool for genome editing [82], later work from Dr. Feng Zhang (Broad Institute) and Dr. George Church applied this system in mammalian cells [81]. Thus, a major controversy has arisen over the patents to this important genome-editing technology as while Dr. Doudna and Dr. Charpentier were first to file, Dr. Zhang and the Broad Institute paid to have their patent application processed in an expedited manner. This case is still being litigated in court with both the University of California-Berkeley and the Broad Institute spending significant resources to litigate this patent [85]. These three researchers have founded start-ups based on this technology with Dr. Doudna being associated with Caribou Biosciences, Dr. Charpentier with CRISPR Therapeutics, and Dr. Zhang with Editas Medicine and these company's websites are given at the end of the chapter.

In addition to the ongoing battle over the patents, there is controversy over whether this technology should be used to edit human embryos due to its off-target effects. Several leading scientists have urged a ban on editing human embryos intended for fertilization, implantation, and being carried to term to produce a child, producing “designer” babies. However, like the previously mentioned technologies for genome editing, CRISPR/Cas9 can be applied to basic biological problems and other therapeutic applications. These tools can alter the genomes of pluripotent stem cells to create disease models or repair the mutations associated with different neurological diseases [86]. Overall, this technology has the power to both address genetic basis of certain neurological diseases and to manipulate stem cell behavior for neural tissue engineering applications. As mentioned in Section 4, such gene editing can be used to directly reprogram cells, providing an additional avenue for neural tissue engineering.

6 OPTOGENETICS AND DESIGNER RECEPTORS EXCLUSIVELY ACTIVATED BY DESIGNER DRUGS (DREADDs)

This section gives a brief overview of how two different types of stimuli (light and chemical) can manipulate the behavior of nerve cells. The first method, optogenetics, relies on the light to stimulate the signaling of neurons [87]. In this technique, target populations of neural cells are virally transfected to express receptors that are sensitive to light, enabling their cellular functions to be controlled when light is applied to the system. This concept of using light to control the nervous system has been around since the 1970s [88]. Recent advances in this technique have made it more accessible to the scientific community and it has been broadly accepted by the neuroscience community as an important tool for elucidating how different neuronal circuits contribute to brain function—both in the healthy and diseased brain.

In 2003, Dr. Miesenböck and his colleagues developed a system for controlling the activity of neurons using light as well as pharmacologically [89] and other groups worked with a different system to achieve similar effects with light [90]. In 2005, the Deisseroth lab published a study that used lentiviral transfection to express the protein channelrhodopsin-2, which is derived from algae and a light-sensitive ion channel [91]. They then showed that these channels were immediately responsive to photostimulation, enabling a high degree control over the signaling of the neurons expressing these channels (Fig. 6). This system has been widely adopted by the neuroscience community as an important tool for elucidating how the nervous system functions. At the same time as this work was being conducted by the Deisseroth lab in part by Dr. Boyden (now at the Massachusetts Institute of Technology), Dr. Zhuo-Hua Pan developed a similar system where he used light to control retinal neurons in particular [92]. This work has now been translated as the first human clinical trial using optogenetics through the company Retrosense. More recently, work has demonstrated how optogenetics can exert functional control over the behavior of human embryonic stem cells in vitro and in vivo [93]. This study also used a fluorescently labeled version of channelrhodopsin-2 to enable tracking of the cells in vivo after transplantation, demonstrating relevance for engineering neural tissue from stem cells as the transplanted cells could be monitored postimplantation. Overall, the use of light to control neural cell behavior enables a better understanding how the nervous system functions and it may lead to innovative treatments for certain neurological disorders as this technology is translated into the clinic. However, it may be challenging to deliver such light stimuli to the human brain.

The second method of using stimuli to manipulate cell behavior works by expressing designer receptors in the target cellular populations that are exclusively activated by designer drugs, which yields the catchy acronym—DREADD [94,95]. In this system, the designer drug replaces the use of light as the trigger for opening the receptor (Fig. 7). Another way of describing this unique technology is the acronym—receptor activated solely by a synthetic ligand (RASSL) [96,97]. In this case, instead of using ion channels that are activated by light, cells are genetically modified to contain synthetic receptors derived by mutating known G-coupled protein receptors that open in response to an artificial ligand, usually clozapine N-oxide (CNO), which is not expressed by humans. Since these receptors only response to this artificial ligand (as opposed to the biomolecules found in the body), it serves as a very selective system for receptor activation, enabling precise activation of the target cellular populations that have been genetically manipulated to express these receptors. Expression of targeted neuronal receptors is achieved by having these synthetic receptors transcribed under their promoters.

Unlike with optogenetics which often requires invasive surgery to stimulate the activity of transfected cells, the synthetic ligands can be administered systemically with minimal invasiveness. Thus, the effects of activating target cellular populations

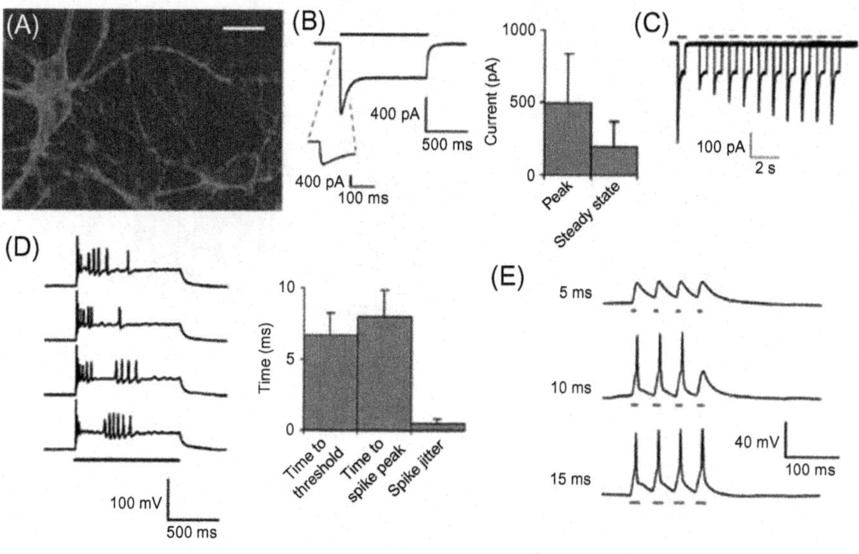

FIG. 6

Optogenetic activation of transfected hippocampal neurons. (A) Hippocampal neurons expressing ChR2-YFP (scale bar 30 μm). (B) Left, inward current in voltage-clamped neuron evoked by 1 s of GFP-wavelength light (indicated by *black bar*); right, population data (right; mean ± s.d. plotted throughout; $n = 18$). Inset, expanded initial phase of the current transient. (C) Ten overlaid current traces recorded from a hippocampal neuron illuminated with pairs of 0.5-s light pulses (indicated by *gray bars*), separated by intervals varying from 1 to 10 s. (D) Voltage traces showing membrane depolarization and spikes in a current-clamped hippocampal neuron (left) evoked by 1-s periods of light *(gray bar)*. Right, properties of the first spike elicited ($n = 10$): latency to spike threshold, latency to spike peak, and jitter of spike time. (E) Voltage traces in response to brief light pulse series, with light pulses *(gray bars)* lasting 5 ms (top), 10 ms (middle), or 15 ms (bottom).

Reprinted with permission from Boyden ES, Zhang F, Bamberg E, Nagel G, Deisseroth K. Millisecond-timescale, genetically targeted optical control of neural activity. Nat Neurosci 2005;8(9):1263–8.

on the resulting behavior can be studied in active model animal systems [98]. DREADD technology can be used to image transfected neural progenitor cells derived from induced pluripotent stem cells in vivo, demonstrating the versatility of this technique [99]. These receptors can also be expressed in germ lines of transgenic mice, furthering the utility of this technology for basic neuroscience applications. It also provides a potentially easier path to clinical translation in comparison to optogenetic technology. The vectors necessary to implement DREADDs experimentally are referenced at the end of the chapter and these reagents are readily available to scientists. Both techniques serve as valuable tools for manipulating cell behavior to enable better understanding of the function of the nervous system as well as providing avenues for potential therapies.

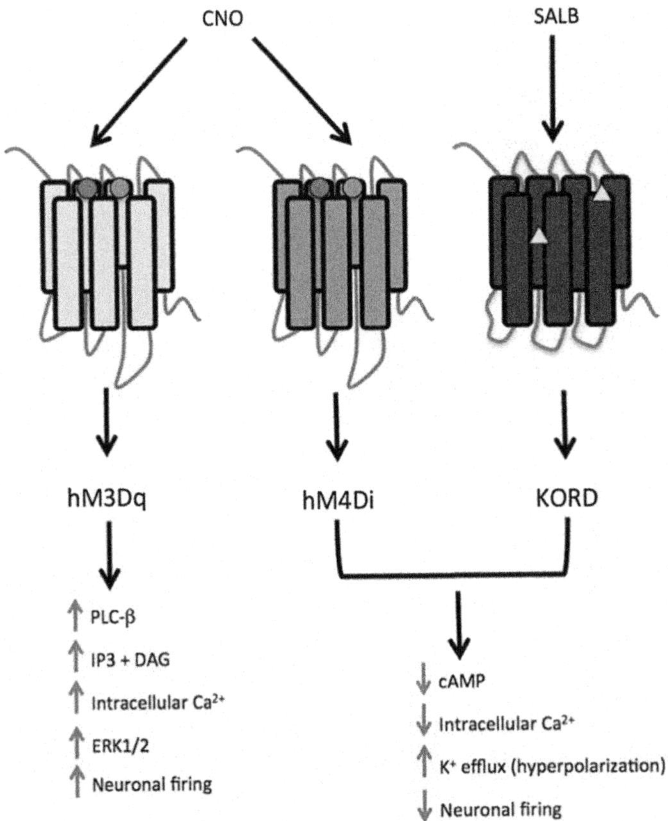

FIG. 7

Summary of modified human muscarinic (hM) and Kappa-opioid receptor (KOR) designer receptors exclusively activated by designer drugs (DREADD) subtypes. hM3Dq, human M3—Gq coupling; hM4Di, human M4—Gi coupling; KORD—Gi coupling. The muscarinic receptors are activated by clozapine N-oxide (CNO), a metabolite of clozapine and the KORD is activated by salvinorin B, an inactive, drug-like metabolite of the KOR-selective agonist salvinorin A (SALA). The intracellular signaling and neuronal activity that results from the activation of these receptors is also depicted. *Circles* represent mutations in TMDIII and TMV for muscarinic receptors whereas *triangles* represent mutations in the "message domain" of TMII and extracellular end of TMVI for KORD.

7 CONCLUSIONS

This chapter has presented a wide variety of novel tools that can be applied to the next generation of engineering neural tissue using stem cells. These tools can also be used to further understanding how the nervous system works and determining what causes different neurological diseases and disorders. In particular, organoids, optogenetics, and DREADDs provide unique ways of studying how the nervous system develops and functions in both healthy and diseased states. On the other hand, 3D bioprinting offers the tantalizing possibility of generating personalized engineered neural tissues tailored to patients based on their genome and condition. Work remains to be done with regards to optimizing bioinks for printing pluripotent stem cells and determining the conditions for differentiating these printed constructs into functional neural tissue. Similarly, the concept of in vivo reprogramming of cells into functional neural tissue holds tremendous promise as potential strategy for regenerating damaged regions of the nervous system. Gene editing offers great potential in terms of correcting neurological diseases at the level of genome. This process can also be used to promote differentiation of stem cells by manipulating gene expression. Finally, the use of DREADDs provide an interesting way to manipulate the behavior of targeted cells populations and it can function as an effective tool for exploring basic mechanisms behind different types of behavior. The ability to use a pharmacological agent to induce changes in neuronal activity also has translational potential for therapeutic applications.

Overall, this book has covered a broad spectrum of information relating to how to engineer neural tissue from stem cells. Chapter 1 introduced the need and demand in terms of the number of people suffering from diseases and disorders of the nervous system before moving onto to a discussion of the anatomical and cellular features of the nervous system in Chapter 2. Chapter 3 focused on the history of stem cells along with the different types of stem cells and their unique properties. Chapter 4 focused on the desirable features to mimic when engineering neural tissue. Then Chapters 5 and 6 discussed how combinations of natural and synthetic biomaterials, respectively, can be combined with stem cells for such applications. Chapter 7 took a closer look at ways of creating controlled drug release for enhancing the ability of such scaffolds to promote the differentiation of stem cells into mature neural tissue. Finally, this chapter covered recent exciting advances that have not yet been full exploited for engineering neural tissue. The technologies discussed in this book have the potential to revolutionize how we treat a range of neurological diseases and disorders.

ADDITIONAL RESOURCES

RESOURCES RELATED TO 3D BIOPRINTING

Aspect Biosystems—Vancouver based 3D bioprinting company: http://aspectbiosystems.com/

Biobots—low cost 3D bioprinting: https://www.biobots.io/

General website on 3D printing: http://3dprinting.com/bio-printing/
Organovo—San Diego based bioprinting company: http://organovo.com/
Se3D bioprinting—resources for educators who want to incorporate 3D printing into the classroom: http://se3d.com/

RESOURCES RELATED TO ORGANOIDS

R&D resource for producing organoids: https://www.rndsystems.com/products/organoid-and-3d-culture-reagents
Cell Press Selections about organoids: http://organoids.elsevierdigitaledition.com/
Organoids Wiki on Wikigenes: https://www.wikigenes.org/e/mesh/e/12370.html

RESOURCES RELATED TO DIRECT REPROGRAMMING OF CELLS

Cell Stem Cell's collection of articles related to direct reprogramming: https://www.wikigenes.org/e/mesh/e/12370.html
The Gladstone Institute's website on their science, including their work on direct reprogramming: https://gladstone.org/u/gctr/index.html
Dr. Marius Wernig's lab at Stanford University: http://stemcellphd.stanford.edu/faculty/marius-wernig.html
Dr. Rudolf Jaenisch's lab at the Massachusetts Institute for Technology: http://wi.mit.edu/people/faculty/jaenisch

RESOURCES RELATED TO GENE EDITING

Nature Methods vide on gene editing nucleases as the 2011 Method of the Year: https://www.youtube.com/watch?v=zDkUFzZoQAs
Dr. Charpentier's lab at Max Planck for Infection Biology: https://www.mpg.de/9343753/infektionsbiologie-charpentier
CRISPR Therapeutics: http://www.crisprtx.com/
Caribou Bioscience: http://cariboubio.com/
Dr. Jennifer Doudna's lab at the University of California-Berkeley: http://rna.berkeley.edu/
Editas Medicine: http://www.editasmedicine.com/
Dr. Feng Zhang's lab at the Massachusetts Institute for Technology: http://zlab.mit.edu/
Sangamo Therapeutics: http://www.sangamo.com/
Sigma-Aldrich's resources related to synthetic biology: http://www.sigmaaldrich.com/life-science/synthetic-biology.html

RESOURCES RELATED TO OPTOGENETICS AND DREDDs

Nature Methods video on optogenetics as the 2010 Method of the Year: https://www.youtube.com/watch?v=I64X7vHSHOE
Resources from Dr. Edward Boyden's lab at the Massachusetts Institute of Technology related to optogenetics: http://syntheticneurobiology.org/protocols

Resources from Dr. Karl Deissroth's lab at Stanford University related to optogenetics: http://web.stanford.edu/group/dlab/optogenetics/index.html

Designer receptors exclusively activated by designer drugs (DREADDs) Wiki: http://pdspit3.mml.unc.edu/projects/dreadd/wiki/WikiStart

Circuit Therapeutics: http://www.circuittx.com/

Retrosense Therapeutics: http://retrosense.com/

REFERENCES

[1] Brickman JM, Serup P. Properties of embryoid bodies. Wiley Interdiscipl Rev Develop Biol 2016;6(2).

[2] Kretzschmar K, Clevers H. Organoids: Modeling development and the stem cell niche in a dish. Developmental Cell 2016;38(6):590–600.

[3] Spence JR, Mayhew CN, Rankin SA, Kuhar MF, Vallance JE, Tolle K, et al. Directed differentiation of human pluripotent stem cells into intestinal tissue in vitro. Nature 2011;470(7332):105–9.

[4] Lancaster MA, Renner M, Martin C-A, Wenzel D, Bicknell LS, Hurles ME, et al. Cerebral organoids model human brain development and microcephaly. Nature 2013;501(7467):373–9.

[5] Lancaster MA, Knoblich JA. Generation of cerebral organoids from human pluripotent stem cells. Nature Protocols 2014;9(10):2329–40.

[6] Chua CW, Shibata M, Lei M, Toivanen R, Barlow LJ, Bergren SK, et al. Single luminal epithelial progenitors can generate prostate organoids in culture. Nature Cell Biology 2014;16(10):951–61.

[7] Camp JG, Badsha F, Florio M, Kanton S, Gerber T, Wilsch-Bräuninger M, et al. Human cerebral organoids recapitulate gene expression programs of fetal neocortex development. Proceedings of the National Academy of Sciences 2015;112(51):15672–7.

[8] Garcez PP, Loiola EC, da Costa RM, Higa LM, Trindade P, Delvecchio R, et al. Zika virus impairs growth in human neurospheres and brain organoids. Science 2016;352(6287):816–8.

[9] Chen HY, Kaya KD, Dong L, Swaroop A. Three-dimensional retinal organoids from mouse pluripotent stem cells mimic in vivo development with enhanced stratification and rod photoreceptor differentiation. Molecular Vision 2016;22:1077.

[10] Eiraku M, Takata N, Ishibashi H, Kawada M, Sakakura E, Okuda S, et al. Self-organizing optic-cup morphogenesis in three-dimensional culture. Nature 2011;472(7341):51–6.

[11] Hartley BJ, Brennand KJ. Neural organoids for disease phenotyping, drug screening and developmental biology studies. Neurochemistry International 2016;S0197-0186(16):30370–9.

[12] Licia B, Tatjana A-Z, Eleonora L, Francesco V, Rosaria CM, da Costa Vasconcelos PF. Emerging Zika virus infection: a rapidly evolving situation. Advances in Experimental Medicine and Biology 2017;972:61–86.

[13] Qian X, Nguyen HN, Song MM, Hadiono C, Ogden SC, Hammack C, et al. Brain-region-specific organoids using mini-bioreactors for modeling ZIKV exposure. Cell 2016;165(5):1238–54.

[14] Cugola FR, Fernandes IR, Russo FB, Freitas BC, Dias JL, Guimarães KP, et al. The Brazilian Zika virus strain causes birth defects in experimental models. Nature 2016;534(7606):267–71.

[15] Dang J, Tiwari SK, Lichinchi G, Qin Y, Patil VS, Eroshkin AM, et al. Zika virus depletes neural progenitors in human cerebral organoids through activation of the innate immune receptor TLR3. Cell Stem Cell 2016;19(2):258–65.
[16] Foster JW, Wahlin K, Adams SM, Birk DE, Zack DJ, Chakravarti S. Cornea organoids from human induced pluripotent stem cells. Scientific Reports 2017;7:41286.
[17] Parfitt DA, Lane A, Ramsden C, Jovanovic K, Coffey PJ, Hardcastle AJ, et al. Using induced pluripotent stem cells to understand retinal ciliopathy disease mechanisms and develop therapies. Biochemical Society Transactions 2016;44(5):1245–51.
[18] Hong N, Yang GH, Lee J, Kim G. 3D bioprinting and its in vivo applications. Journal of Biomedical Materials Research Part B: Applied Biomaterials 2017.
[19] Panwar A, Tan LP. Current status of Bioinks for micro-extrusion-based 3D Bioprinting. Molecules 2016;21(6):685.
[20] Xu T, Gregory CA, Molnar P, Cui X, Jalota S, Bhaduri SB, et al. Viability and electrophysiology of neural cell structures generated by the inkjet printing method. Biomaterials 2006;27(19):3580–8.
[21] Lee W, Pinckney J, Lee V, Lee J-H, Fischer K, Polio S, et al. Three-dimensional bioprinting of rat embryonic neural cells. NeuroReport 2009;20(8):798–803.
[22] Lee Y-B, Polio S, Lee W, Dai G, Menon L, Carroll RS, et al. Bio-printing of collagen and VEGF-releasing fibrin gel scaffolds for neural stem cell culture. Experimental Neurology 2010;223(2):645–52.
[23] Hinton TJ, Jallerat Q, Palchesko RN, Park JH, Grodzicki MS, Shue H-J, et al. Three-dimensional printing of complex biological structures by freeform reversible embedding of suspended hydrogels. Science Advances 2015;1(9):e1500758.
[24] Wüst S, Müller R, Hofmann S. 3D Bioprinting of complex channels—Effects of material, orientation, geometry, and cell embedding. Journal of Biomedical Materials Research Part A 2015;103(8):2558–70.
[25] Pateman CJ, Harding AJ, Glen A, Taylor CS, Christmas CR, Robinson PP, et al. Nerve guides manufactured from photocurable polymers to aid peripheral nerve repair. Biomaterials 2015;49:77–89.
[26] Zhu S, Zhu Q, Liu X, Yang W, Jian Y, Zhou X, et al. Three-dimensional reconstruction of the microstructure of human acellular nerve allograft. Scientific Reports 2016;6.
[27] Hu Y, Wu Y, Gou Z, Tao J, Zhang J, Liu Q, et al. 3D-engineering of cellularized conduits for peripheral nerve regeneration. Scientific Reports 2016;6.
[28] Rajaram A, Chen X-B, Schreyer DJ. Strategic design and recent fabrication techniques for bioengineered tissue scaffolds to improve peripheral nerve regeneration. Tissue Engineering Part B: Reviews 2012;18(6):454–67.
[29] Lozano R, Stevens L, Thompson BC, Gilmore KJ, Gorkin R, Stewart EM, et al. 3D printing of layered brain-like structures using peptide modified gellan gum substrates. Biomaterials 2015;67:264–73.
[30] Hsieh F-Y, Hsu S-h. 3D bioprinting: a new insight into the therapeutic strategy of neural tissue regeneration. Organogenesis 2015;11(4):153–8.
[31] Gu Q, Tomaskovic-Crook E, Lozano R, Chen Y, Kapsa RM, Zhou Q, et al. Functional 3D neural mini-tissues from printed gel-based Bioink and human neural stem cells. Advanced Healthcare Materials 2016;5(12):1429–38.
[32] Müller M, Becher J, Schnabelrauch M, Zenobi-Wong M. Printing thermoresponsive reverse molds for the creation of patterned two-component hydrogels for 3D cell culture. Journal of Visualized Experiments 2013;77.

[33] Bsoul A, Pan S, Cretu E, Stoeber B, Walus K. Design, microfabrication, and characterization of a moulded PDMS/SU-8 inkjet dispenser for a lab-on-a-printer platform technology with disposable microfluidic chip. Lab on a Chip 2016;16(17):3351–61.

[34] Takahashi K, Yamanaka S. Induction of pluripotent stem cells from mouse embryonic and adult fibroblast cultures by defined factors. Cell 2006;126(4):663–76.

[35] Takahashi K, Tanabe K, Ohnuki M, Narita M, Ichisaka T, Tomoda K, et al. Induction of pluripotent stem cells from adult human fibroblasts by defined factors. Cell 2007;131(5):861–72.

[36] Okita K, Ichisaka T, Yamanaka S. Generation of germline-competent induced pluripotent stem cells. Nature 2007;448(7151):313–7.

[37] Vierbuchen T, Ostermeier A, Pang ZP, Kokubu Y, Sudhof TC, Wernig M. Direct conversion of fibroblasts to functional neurons by defined factors. Nature 2010;463(7284):1035–41.

[38] Pang ZPP, Yang N, Vierbuchen T, Ostermeier A, Fuentes DR, Yang TQ, et al. Induction of human neuronal cells by defined transcription factors. Nature 2011;476(7359):220-U122.

[39] Marro S, Pang ZP, Yang N, Tsai MC, Qu K, Chang HY, et al. Direct lineage conversion of terminally differentiated hepatocytes to functional neurons. Cell Stem Cell 2011;9(4):374–82.

[40] Yang N, Ng YH, Pang ZP, Sudhof TC, Wernig M. Induced neuronal cells: how to make and define a neuron. Cell Stem Cell 2011;9(6):517–25.

[41] Chanda S, Marro S, Wernig M, Sudhof TC. Neurons generated by direct conversion of fibroblasts reproduce synaptic phenotype caused by autism-associated neuroligin-3 mutation. Proc Natl Acad Sci U S A 2013;110(41):16622–7.

[42] Liu M-L, Zang T, Zhang C-L. Direct lineage reprogramming reveals disease-specific phenotypes of motor neurons from human ALS patients. Cell Reports 2016;14(1):115–28.

[43] Huang Y, Tan S. Direct lineage conversion of astrocytes to induced neural stem cells or neurons. Neuroscience Bulletin 2015;31(3):357–67.

[44] Lujan E, Chanda S, Ahlenius H, Sudhof TC, Wernig M. Direct conversion of mouse fibroblasts to self-renewing, tripotent neural precursor cells. Proc Natl Acad Sci U S A 2012;109(7):2527–32.

[45] Chanda S, Ang CE, Davila J, Pak C, Mall M, Lee QY, et al. Generation of induced neuronal cells by the single reprogramming factor ASCL1. Stem Cell Reports 2014;3(2):282–96.

[46] Niu W, Zang T, Zou Y, Fang S, Smith DK, Bachoo R, et al. In vivo reprogramming of astrocytes to neuroblasts in the adult brain. Nat Cell Biol 2013;15(10):1164–75.

[47] Niu W, Zang T, Smith DK, Vue TY, Zou Y, Bachoo R, et al. SOX2 reprograms resident astrocytes into neural progenitors in the adult brain. Stem Cell Reports 2015;4(5):780–94.

[48] Vasconcelos FF, Castro DS. Transcriptional control of vertebrate neurogenesis by the proneural factor Ascl1. Front Cell Neurosci 2014;8:412.

[49] Wapinski OL, Vierbuchen T, Qu K, Lee QY, Chanda S, Fuentes DR, et al. Hierarchical mechanisms for direct reprogramming of fibroblasts to neurons. Cell 2013;155(3):621–35.

[50] Caiazzo M, Dell'Anno MT, Dvoretskova E, Lazarevic D, Taverna S, Leo D, et al. Direct generation of functional dopaminergic neurons from mouse and human fibroblasts. Nature 2011;476(7359):224–7.

[51] Xue Y, Qian H, Hu J, Zhou B, Zhou Y, Hu X, et al. Sequential regulatory loops as key gatekeepers for neuronal reprogramming in human cells. Nature Neuroscience 2016;19(6):807–15.

[52] Richner M, Victor MB, Liu Y, Abernathy D, Yoo AS. MicroRNA-based conversion of human fibroblasts into striatal medium spiny neurons. Nature Protocols 2015;10(10):1543–55.

[53] Zhou C, Gu H, Fan R, Wang B, Lou J. MicroRNA 302/367 cluster effectively facilitates direct reprogramming from human fibroblasts into functional neurons. Stem Cells and Development 2015;24(23):2746–55.

[54] Meyer S, Wörsdörfer P, Günther K, Thier M, Edenhofer F. Derivation of adult human fibroblasts and their direct conversion into expandable neural progenitor cells. Journal of Visualized Experiments 2015;(101)e52831-e.

[55] Black JB, Adler AF, Wang H-G, D'Ippolito AM, Hutchinson HA, Reddy TE, et al. Targeted epigenetic remodeling of endogenous loci by CRISPR/Cas9-based transcriptional activators directly converts fibroblasts to neuronal cells. Cell Stem Cell 2016;19(3):406–14.

[56] Han Y-C, Lim Y, Duffieldl MD, Li H, Liu J, Abdul Manaph NP, et al. Direct reprogramming of mouse fibroblasts to neural stem cells by small molecules. Stem Cells International 2015;2016.

[57] Li X, Zuo X, Jing J, Ma Y, Wang J, Liu D, et al. Small-molecule-driven direct reprogramming of mouse fibroblasts into functional neurons. Cell Stem Cell 2015;17(2):195–203.

[58] Hu W, Qiu B, Guan W, Wang Q, Wang M, Li W, et al. Direct conversion of normal and Alzheimer's disease human fibroblasts into neuronal cells by small molecules. Cell Stem Cell 2015;17(2):204–12.

[59] Zheng J, Choi K-A, Kang PJ, Hyeon S, Kwon S, Moon J-H, et al. A combination of small molecules directly reprograms mouse fibroblasts into neural stem cells. Biochemical and Biophysical Research Communications 2016;476(1):42–8.

[60] Tian E, Sun G, Sun G, Chao J, Ye P, Warden C, et al. Small-molecule-based lineage reprogramming creates functional astrocytes. Cell Reports 2016;16(3):781–92.

[61] Srivastava D, DeWitt N. In vivo cellular reprogramming: The next generation. Cell 2016;166(6):1386–96.

[62] Inagawa K, Miyamoto K, Yamakawa H, Muraoka N, Sadahiro T, Umei T, et al. Induction of cardiomyocyte-like cells in infarct hearts by gene transfer of Gata4, Mef2c, and Tbx5. Circulation Research 2012;111(9):1147–56.

[63] Ma H, Wang L, Yin C, Liu J, Qian L. In vivo cardiac reprogramming using an optimal single polycistronic construct. Cardiovascular Research 2015;108(2):217–9.

[64] Guo Z, Zhang L, Wu Z, Chen Y, Wang F, Chen G. In vivo direct reprogramming of reactive glial cells into functional neurons after brain injury and in an Alzheimer's disease model. Cell Stem Cell 2014;14(2):188–202.

[65] Torper O, Pfisterer U, Wolf DA, Pereira M, Lau S, Jakobsson J, et al. Generation of induced neurons via direct conversion in vivo. Proceedings of the National Academy of Sciences 2013;110(17):7038–43.

[66] Torper O, Ottosson DR, Pereira M, Lau S, Cardoso T, Grealish S, et al. In vivo reprogramming of striatal NG2 glia into functional neurons that integrate into local host circuitry. Cell Reports 2015;12(3):474–81.

[67] Liu Y, Miao Q, Yuan J, Han S, Zhang P, Li S, et al. Ascl1 converts dorsal midbrain astrocytes into functional neurons in vivo. J Neurosci 2015;35(25):9336–55.

[68] Su ZD, Niu WZ, Liu ML, Zou YH, Zhang CL. In vivo conversion of astrocytes to neurons in the injured adult spinal cord. Nature Communications 2014;5.

[69] Serguera C, Bemelmans AP. Gene therapy of the central nervous system: general considerations on viral vectors for gene transfer into the brain. Rev Neurol (Paris) 2014;170(12):727–38.

[70] Treutlein B, Lee QY, Camp JG, Mall M, Koh W, Shariati SAM, et al. Dissecting direct reprogramming from fibroblast to neuron using single-cell RNA-seq. Nature 2016;534(7607):391–5.

[71] Di Stefano B, Hochedlinger K. Cell reprogramming: brain versus brawn. Nature 2016;534(7607):332–3.
[72] Nawy T. Stem cells: fast track to neurons. Nat Methods 2015;12(10):915.
[73] Robinson M, Chapani P, Styan T, Vaidyanathan R, Willerth SM. Functionalizing Ascl1 with novel intracellular protein delivery Technology for promoting neuronal differentiation of human induced pluripotent stem cells. Stem Cell Rev Rep 2016;1–8.
[74] Willerth SM. Using functionalized transcription factors to engineer personalized neural tissue. Neural Regeneration Research 2016;11(10):1570–1.
[75] Gaj T, Gersbach CA, Barbas CF. ZFN, TALEN, and CRISPR/Cas-based methods for genome engineering. Trends in Biotechnology 2013;31(7):397–405.
[76] McMahon MA, Cleveland DW. Gene therapy: gene-editing therapy for neurological disease. Nature Reviews Neurology 2016;13(1):7–9.
[77] Kim Y-G, Cha J, Chandrasegaran S. Hybrid restriction enzymes: zinc finger fusions to Fok I cleavage domain. Proceedings of the National Academy of Sciences 1996;93(3):1156–60.
[78] Tebas P, Stein D, Tang WW, Frank I, Wang SQ, Lee G, et al. Gene editing of CCR5 in autologous CD4 T cells of persons infected with HIV. New England Journal of Medicine 2014;370(10):901–10.
[79] Garriga-Canut M, Agustín-Pavón C, Herrmann F, Sánchez A, Dierssen M, Fillat C, et al. Synthetic zinc finger repressors reduce mutant huntingtin expression in the brain of R6/2 mice. Proceedings of the National Academy of Sciences 2012;109(45):E3136–45.
[80] Joung JK, Sander JD. TALENs: a widely applicable technology for targeted genome editing. Nature Reviews Molecular Cell Biology 2013;14(1):49–55.
[81] Zhang F, Cong L, Lodato S, Kosuri S, Church GM, Arlotta P. Efficient construction of sequence-specific TAL effectors for modulating mammalian transcription. Nature Biotechnology 2011;29(2):149–53.
[82] Jinek M, Chylinski K, Fonfara I, Hauer M, Doudna JA, Charpentier E. A programmable dual-RNA–guided DNA endonuclease in adaptive bacterial immunity. Science 2012;337(6096):816–21.
[83] Mali P, Yang L, Esvelt KM, Aach J, Guell M, DiCarlo JE, et al. RNA-guided human genome engineering via Cas9. Science 2013;339(6121):823–6.
[84] Cong L, Ran FA, Cox D, Lin S, Barretto R, Habib N, et al. Multiplex genome engineering using CRISPR/Cas systems. Science 2013;339(6121):819–23.
[85] Reardon S. Crispr heavyweights battle in US patent court. Nature 2016;540(7633):326–7.
[86] Gerlai R. Gene targeting using homologous recombination in embryonic stem cells: the future for behavior genetics? Frontiers in Genetics 2016;7.
[87] Pathak GP, Vrana JD, Tucker CL. Optogenetic control of cell function using engineered photoreceptors. Biology of the Cell 2013;105(2):59–72.
[88] Fork RL. Laser stimulation of nerve cells in Aplysia. Science 1971;171(3974):907–8.
[89] Zemelman BV, Nesnas N, Lee GA, Miesenböck G. Photochemical gating of heterologous ion channels: remote control over genetically designated populations of neurons. Proceedings of the National Academy of Sciences 2003;100(3):1352–7.
[90] Zemelman BV, Lee GA, Ng M, Miesenböck G. Selective photostimulation of genetically chARGed neurons. Neuron 2002;33(1):15–22.
[91] Boyden ES, Zhang F, Bamberg E, Nagel G, Deisseroth K. Millisecond-timescale, genetically targeted optical control of neural activity. Nature Neuroscience 2005;8(9):1263–8.
[92] Bi A, Cui J, Ma Y-P, Olshevskaya E, Pu M, Dizhoor AM, et al. Ectopic expression of a microbial-type rhodopsin restores visual responses in mice with photoreceptor degeneration. Neuron 2006;50(1):23–33.

[93] Weick JP, Johnson MA, Skroch SP, Williams JC, Deisseroth K, Zhang SC. Functional control of transplantable human ESC-derived neurons via Optogenetic targeting. Stem Cells 2010;28(11):2008–16.
[94] Roth BL. DREADDs for neuroscientists. Neuron 2016;89(4):683–94.
[95] Urban DJ, Roth BL. DREADDs (designer receptors exclusively activated by designer drugs): chemogenetic tools with therapeutic utility. Annual Review of Pharmacology and Toxicology 2015;55:399–417.
[96] Armbruster BN, Li X, Pausch MH, Herlitze S, Roth BL. Evolving the lock to fit the key to create a family of G protein-coupled receptors potently activated by an inert ligand. Proceedings of the National Academy of Sciences 2007;104(12):5163–8.
[97] Conklin BR, Hsiao EC, Claeysen S, Dumuis A, Srinivasan S, Forsayeth JR, et al. Engineering GPCR signaling pathways with RASSLs. Nature Methods 2008;5(8):673–8.
[98] Ferguson SM, Neumaier JF. Grateful DREADDs: engineered receptors reveal how neural circuits regulate behavior. Neuropsychopharmacology 2012;37(1):296.
[99] Ji B, Kaneko H, Minamimoto T, Inoue H, Takeuchi H, Kumata K, et al. Multimodal imaging for DREADD-expressing neurons in living brain and their application to implantation of iPSC-derived neural progenitors. Journal of Neuroscience 2016;36(45):11544–58.

Index

Note: Page numbers followed by *f* indicate figures.

K

L

M

N

V

W

Z